Atlas of the Human Body

Atlas of the Human Body
Central Nervous System and Vascularization

BRANISLAV VIDIĆ, DS
Professor, Texas Tech University Health
Science Center, Lubbock, TX,
United States

MILAN MILISAVLJEVIĆ, MD, DS, DSc
Professor, Institute of Anatomy,
University of Belgrade, Belgrade, Serbia

in collaboration with

ALEKSANDAR MALIKOVIĆ, MD, DSc
Professor, Institute of Anatomy,
University of Belgrade, Belgrade, Serbia

ACADEMIC PRESS
An imprint of Elsevier

Academic Press is an imprint of Elsevier
125 London Wall, London EC2Y 5AS, United Kingdom
525 B Street, Suite 1800, San Diego, CA 92101-4495, United States
50 Hampshire Street, 5th Floor, Cambridge, MA 02139, United States
The Boulevard, Langford Lane, Kidlington, Oxford OX5 1GB, United Kingdom

Notices
Knowledge and best practice in this field are constantly changing. As new research and experience broaden our understanding, changes in research methods, professional practices, or medical treatment may become necessary.

Practitioners and researchers must always rely on their own experience and knowledge in evaluating and using any information, methods, compounds, or experiments described herein. In using such information or methods they should be mindful of their own safety and the safety of others, including parties for whom they have a professional responsibility.

To the fullest extent of the law, neither the Publisher nor the authors, contributors, or editors, assume any liability for any injury and/or damage to persons or property as a matter of products liability, negligence or otherwise, or from any use or operation of any methods, products, instructions, or ideas contained in the material herein.

Library of Congress Cataloging-in-Publication Data
A catalog record for this book is available from the Library of Congress

British Library Cataloguing-in-Publication Data
A catalogue record for this book is available from the British Library

ISBN: 978-0-12-809410-5

For information on all Academic Press publications visit our website at
https://www.elsevier.com/books-and-journals

Working together
to grow libraries in
developing countries

www.elsevier.com • www.bookaid.org

Publisher: Mica Haley
Acquisition Editor: Stacy Masucci
Editorial Project Manager: Sam Young
Production Project Manager: Edward Taylor
Designer: Matthew Limbert

Typeset by Thomson Digital

Contents

Preface

Anatomy is one of the oldest medical sciences that still continues today, and is the foundation for the study and practice of the medical arts. It provides, first of all, the basic vocabulary of the medical world and the necessary skills required in solving health–disease problems in three-dimensional space. By sequentially dissecting a region, the anatomical analysis leads to a gradually expanding appreciation of the entire makeup of the human body. This process is fundamental in providing biophysical data for subsequent conceptual elaboration and integration of morphological data into a meaningful functional complex. Dissection, considered the most ancient method of studying an anatomical subject, survived scrutiny and the test of time throughout the history of medicine. However, it still remains a reliable method of scientific and pedagogical analysis of fundamental human structure and function, which is important for minute differential assessment between normal and abnormal conditions, as well as for the optimal treatment of abnormal (diseased) conditions.

Atlas of the Human Body, Central Nervous System and Vascularization has been written with several goals in mind. The most important one was to establish a detailed coverage of anatomical structures/relationships throughout topographic regions, as completely as it was technically possible. To avoid overcrowding of photographs by labels and to provide better visibility of images, two, three, or even more similar regional views, in some instances, have been utilized. In addition to adult specimens, a few prenatal examples were utilized to enable a better understanding of structure/relation specificity of corporal differentiation (conduits, organs, somatic, and branchial derivatives) at various developmental intervals. Another quest of this Atlas was to systematically present arterial distribution, up to the precapillary level, using the "methyl methacrylate injection and subsequent digestion of tissue" method. The resulting photographic presentation of the arterial distribution throughout topographic regions, organs, and special subregions makes this Atlas a unique and invaluable published document in the arsenal of the existing academic literature. The present Atlas, furthermore, contains a very rich collection of: surface and three-dimensional dissection images, native and colored cross-sectional views made in different plans (whenever appropriate these views were compared, side by side, with dissection images), and the distribution of blood vessels throughout body regions and central nervous system. A separate segment of Atlas is devoted to the central nervous system and its specific regions: brain, brainstem, cerebellum, and spinal cord. Each region is presented by a detailed collection of surface (dissection) and cross-sectional views, native blood vessels, and blood vessel casts. The latter collection could adequately subserve as a complete educational–visual aid for the requirements of a Neuroscience course.

Terminology used in the *Atlas of the Human Body, Central Nervous System, and Vascularization* is according to the Terminologia Anatomica (1998).

Authors express their sincere gratitude to Stacy Masucci, Senior Acquisition Editor, Elsevier Inc. and Samuel Young, Editorial Project Manager, Elsevier Inc. for unlimited assistance and help during the course of preparation of *Atlas of the Human Body, Central Nervous System and Vascularization*. Administrative help and encouragement from the Faculty of Medicine, University of Belgrade, Belgrade, Serbia, and Texas Tech University Health Science Center, Texas Tech University, Lubbock, Texas, USA are well appreciated.

We are also deeply grateful for essential scientific contributions by:

Dr. Mila Ćetković Milisavljević, from the Institute of Histology and Embryology, Faculty of Medicine, University of Belgrade, Serbia, for her quality, beautiful drawings and histological specimens used in this Atlas.

Dr. Zdravko Vitošević, University Professor and 2011 year Laureate of the "Brothers Karic Foundation" in Belgrade, Serbia, was wholly committed to this project and helped in the organization of scientific material throughout.

Dr. Bojan Štimec, from the Faculty of Medicine, Department of Cellular Physiology and Metabolism, Anatomy Sector, University of Geneva, Geneva, Switzerland, who provided numerous helpful suggestions and assisted in updating the lower limb and abdomen parts of our Atlas.

Chapter 1

Upper Limb and Vascularization

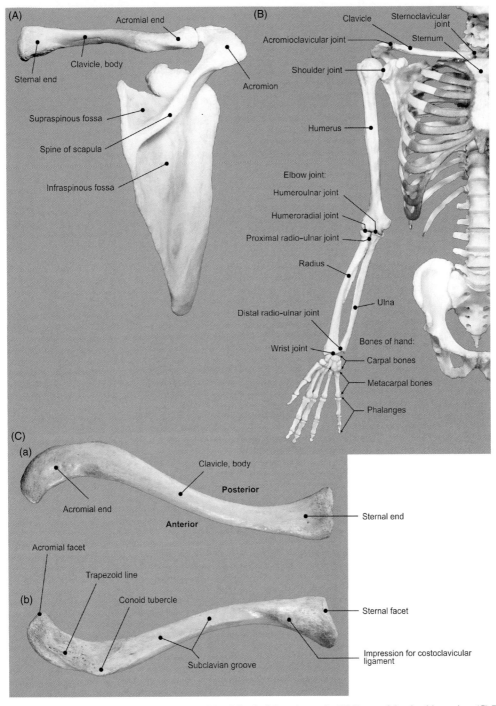

FIGURE 1.1 Skeleton of the upper limb. (A) Posterior view of the right clavicle and scapula. (B) Bones of the shoulder region. (C) Right clavicle: (a) superior and (b) inferior views.

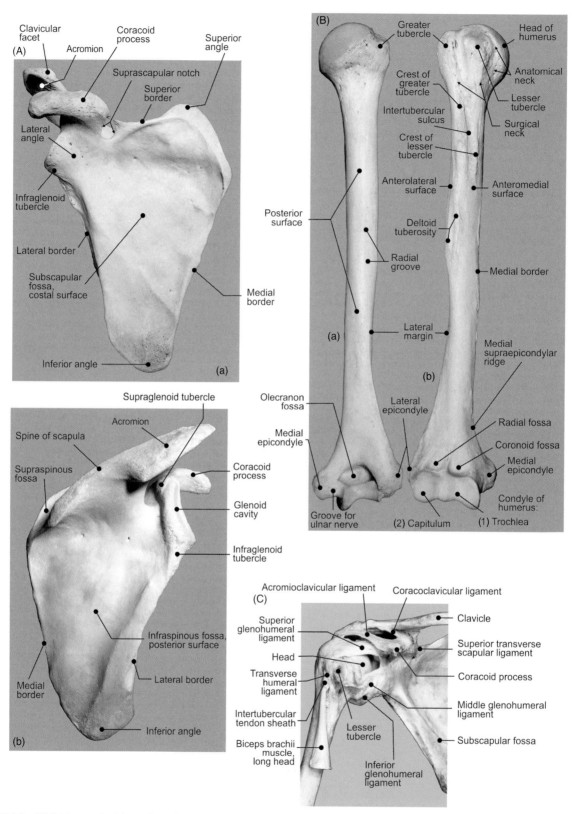

FIGURE 1.2 (A) Right scapula: (a) anterior and (b) posterior views. (B) Right humerus: (a) posterior and (b) anterior views. (C) Anterior view of the shoulder joint.

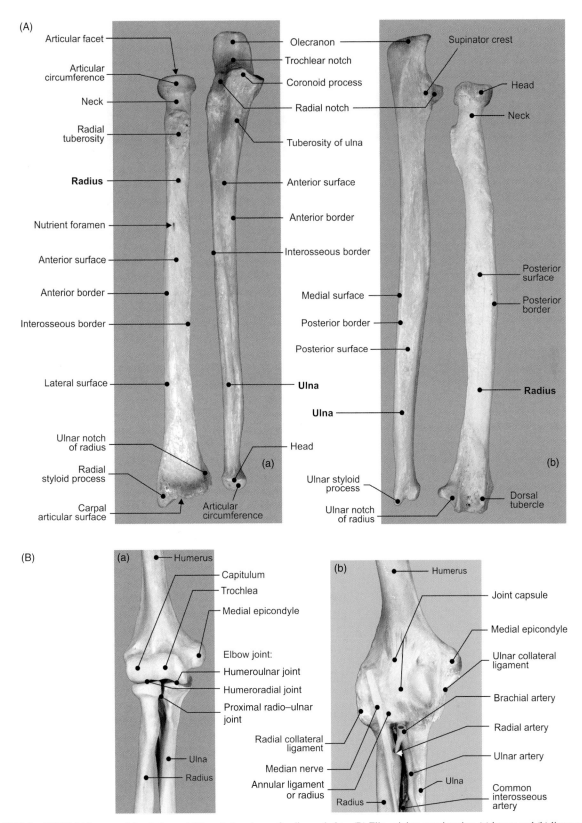

FIGURE 1.3 (A) Right forearm: (a) anterior and (b) posterior views of radius and ulna. (B) Elbow joint: anterior view (a) bones and (b) ligaments.

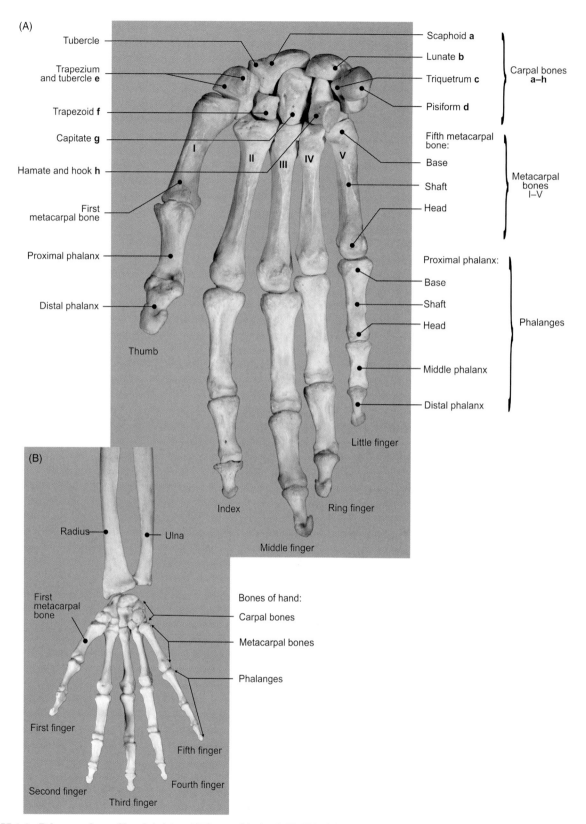

(A)

Tubercle

Trapezium and tubercle **e**

Trapezoid **f**

Capitate **g**

Hamate and hook **h**

First metacarpal bone

Proximal phalanx

Distal phalanx

Thumb

Scaphoid **a**

Lunate **b**

Triquetrum **c**

Pisiform **d**

Carpal bones **a–h**

Fifth metacarpal bone:

Base

Shaft

Head

Metacarpal bones I–V

Proximal phalanx:

Base

Shaft

Head

Middle phalanx

Distal phalanx

Phalanges

I   II   III   IV   V

Little finger

Index

Middle finger

Ring finger

(B)

Radius

Ulna

First metacarpal bone

First finger

Second finger

Third finger

Fourth finger

Fifth finger

Bones of hand:

Carpal bones

Metacarpal bones

Phalanges

FIGURE 1.4 **Palmar surfaces of hand skeleton.** (A) Bones of the hand. (B) Wrist joint.

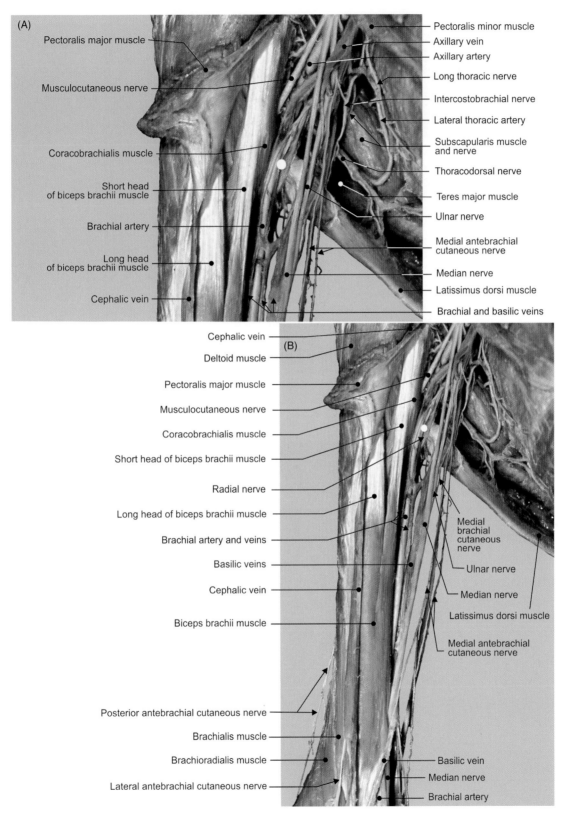

FIGURE 1.5   Anterior views of the (A) axillary and (B) brachial regions.

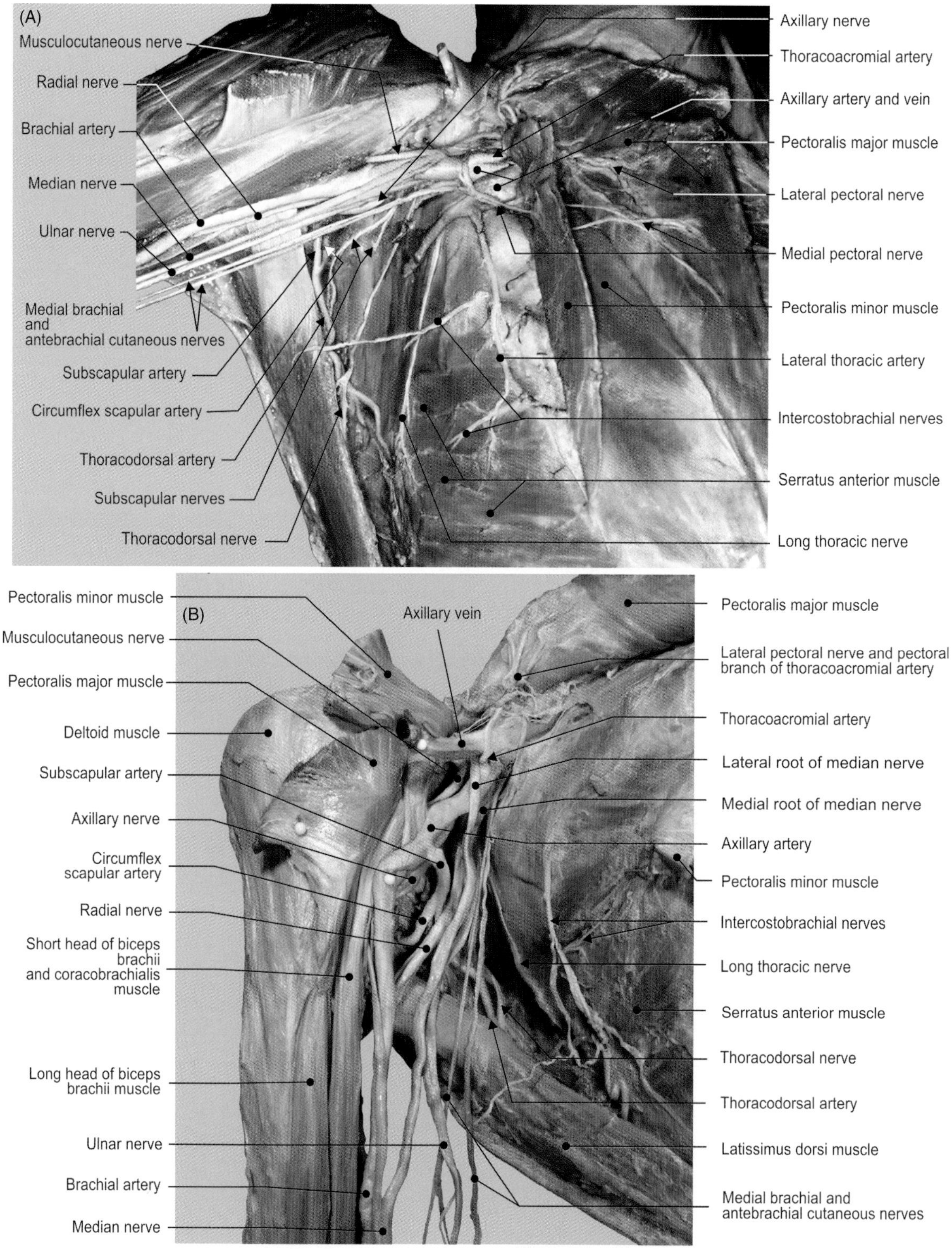

(A)

Musculocutaneous nerve

Radial nerve

Brachial artery

Median nerve

Ulnar nerve

Medial brachial and antebrachial cutaneous nerves

Subscapular artery

Circumflex scapular artery

Thoracodorsal artery

Subscapular nerves

Thoracodorsal nerve

Axillary nerve

Thoracoacromial artery

Axillary artery and vein

Pectoralis major muscle

Lateral pectoral nerve

Medial pectoral nerve

Pectoralis minor muscle

Lateral thoracic artery

Intercostobrachial nerves

Serratus anterior muscle

Long thoracic nerve

(B)

Pectoralis minor muscle

Musculocutaneous nerve

Pectoralis major muscle

Deltoid muscle

Subscapular artery

Axillary nerve

Circumflex scapular artery

Radial nerve

Short head of biceps brachii and coracobrachialis muscle

Long head of biceps brachii muscle

Ulnar nerve

Brachial artery

Median nerve

Axillary vein

Pectoralis major muscle

Lateral pectoral nerve and pectoral branch of thoracoacromial artery

Thoracoacromial artery

Lateral root of median nerve

Medial root of median nerve

Axillary artery

Pectoralis minor muscle

Intercostobrachial nerves

Long thoracic nerve

Serratus anterior muscle

Thoracodorsal nerve

Thoracodorsal artery

Latissimus dorsi muscle

Medial brachial and antebrachial cutaneous nerves

FIGURE 1.6   Anterior views of the (A) axillary and (B) brachial regions.

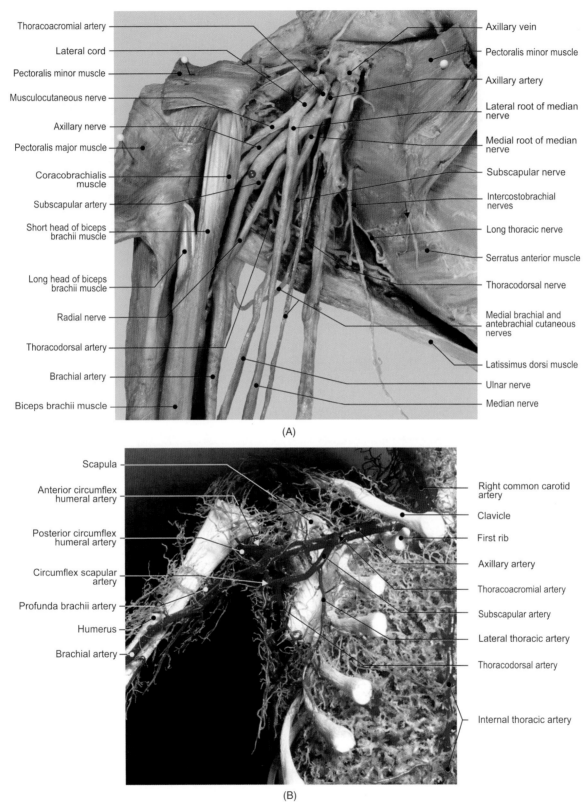

Thoracoacromial artery

Lateral cord

Pectoralis minor muscle

Musculocutaneous nerve

Axillary nerve

Pectoralis major muscle

Coracobrachialis muscle

Subscapular artery

Short head of biceps brachii muscle

Long head of biceps brachii muscle

Radial nerve

Thoracodorsal artery

Brachial artery

Biceps brachii muscle

Axillary vein

Pectoralis minor muscle

Axillary artery

Lateral root of median nerve

Medial root of median nerve

Subscapular nerve

Intercostobrachial nerves

Long thoracic nerve

Serratus anterior muscle

Thoracodorsal nerve

Medial brachial and antebrachial cutaneous nerves

Latissimus dorsi muscle

Ulnar nerve

Median nerve

(A)

Scapula

Anterior circumflex humeral artery

Posterior circumflex humeral artery

Circumflex scapular artery

Profunda brachii artery

Humerus

Brachial artery

Right common carotid artery

Clavicle

First rib

Axillary artery

Thoracoacromial artery

Subscapular artery

Lateral thoracic artery

Thoracodorsal artery

Internal thoracic artery

(B)

FIGURE 1.7 (A) Anterior view of the axillary fossa. (B) Anterior view of the fetal arterial distribution over rib cage and arm (corrosion cast).

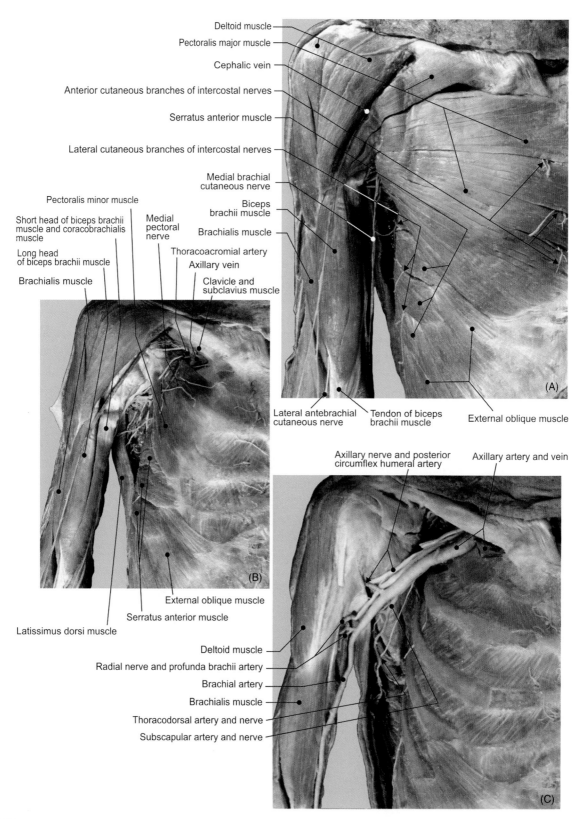

Deltoid muscle

Pectoralis major muscle

Cephalic vein

Anterior cutaneous branches of intercostal nerves

Serratus anterior muscle

Lateral cutaneous branches of intercostal nerves

Medial brachial cutaneous nerve

Biceps brachii muscle

Brachialis muscle

Thoracoacromial artery

Axillary vein

Clavicle and subclavius muscle

Pectoralis minor muscle

Short head of biceps brachii muscle and coracobrachialis muscle

Long head of biceps brachii muscle

Brachialis muscle

Medial pectoral nerve

Lateral antebrachial cutaneous nerve

Tendon of biceps brachii muscle

External oblique muscle

External oblique muscle

Serratus anterior muscle

Latissimus dorsi muscle

Axillary nerve and posterior circumflex humeral artery

Axillary artery and vein

Deltoid muscle

Radial nerve and profunda brachii artery

Brachial artery

Brachialis muscle

Thoracodorsal artery and nerve

Subscapular artery and nerve

(A)

(B)

(C)

FIGURE 1.8 **The pectoral region.** (A) Superficial layer. (B) Deep layer. (C) Axillary fossa after the removal of medial and lateral cords of the brachial plexus.

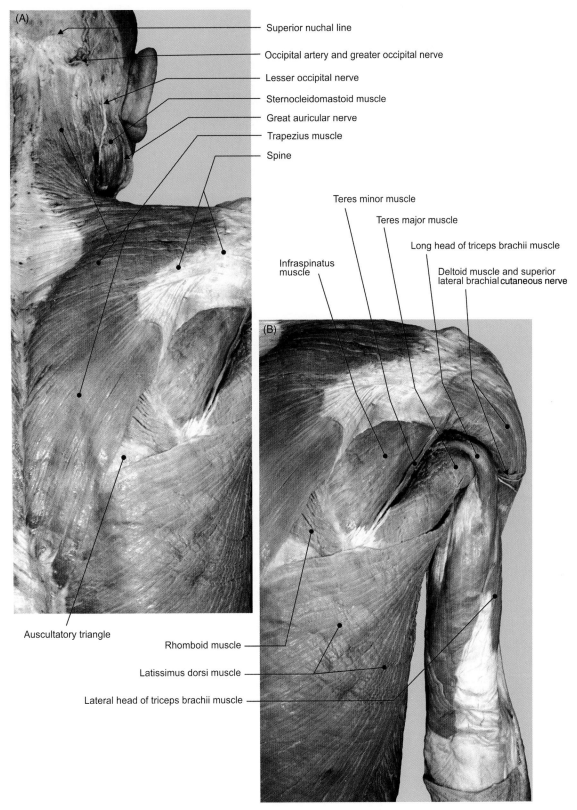

FIGURE 1.9   Superficial layer of the scapular region. (A) Scapular and (B) posterior brachial regions.

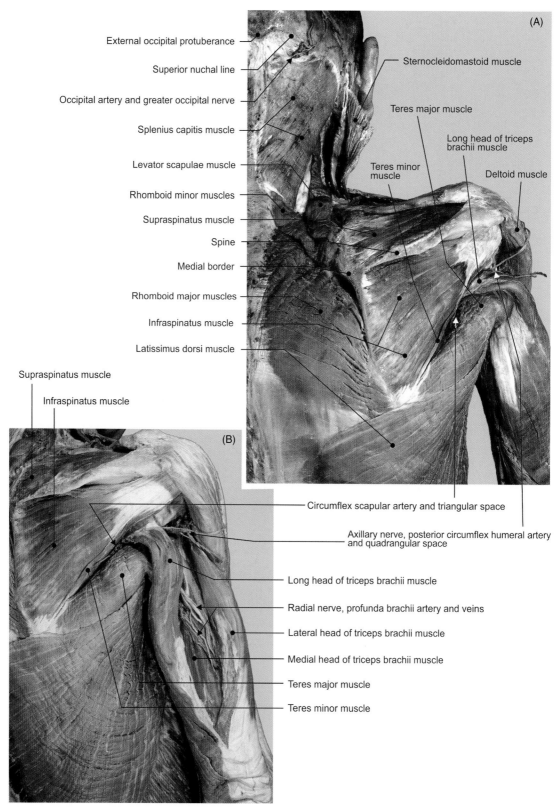

External occipital protuberance

Superior nuchal line

Occipital artery and greater occipital nerve

Splenius capitis muscle

Levator scapulae muscle

Rhomboid minor muscles

Supraspinatus muscle

Spine

Medial border

Rhomboid major muscles

Infraspinatus muscle

Latissimus dorsi muscle

Sternocleidomastoid muscle

Teres major muscle

Long head of triceps brachii muscle

Teres minor muscle

Deltoid muscle

(A)

Supraspinatus muscle

Infraspinatus muscle

(B)

Circumflex scapular artery and triangular space

Axillary nerve, posterior circumflex humeral artery and quadrangular space

Long head of triceps brachii muscle

Radial nerve, profunda brachii artery and veins

Lateral head of triceps brachii muscle

Medial head of triceps brachii muscle

Teres major muscle

Teres minor muscle

FIGURE 1.10 (A) Middle layer of the scapular region and (B) the posterior brachial region.

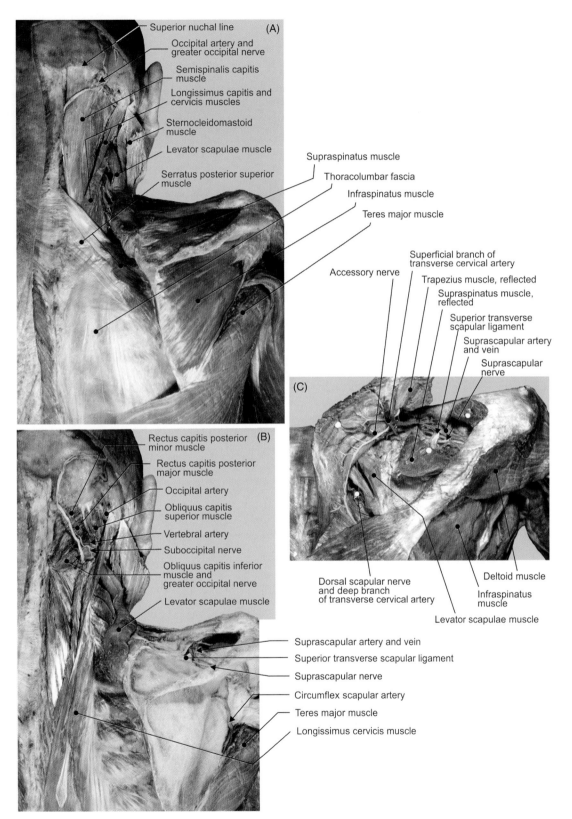

FIGURE 1.11 **Deep layer of the scapular region.** Illustrations of structures: (A) muscles, (B) vascularization, (C) and details.

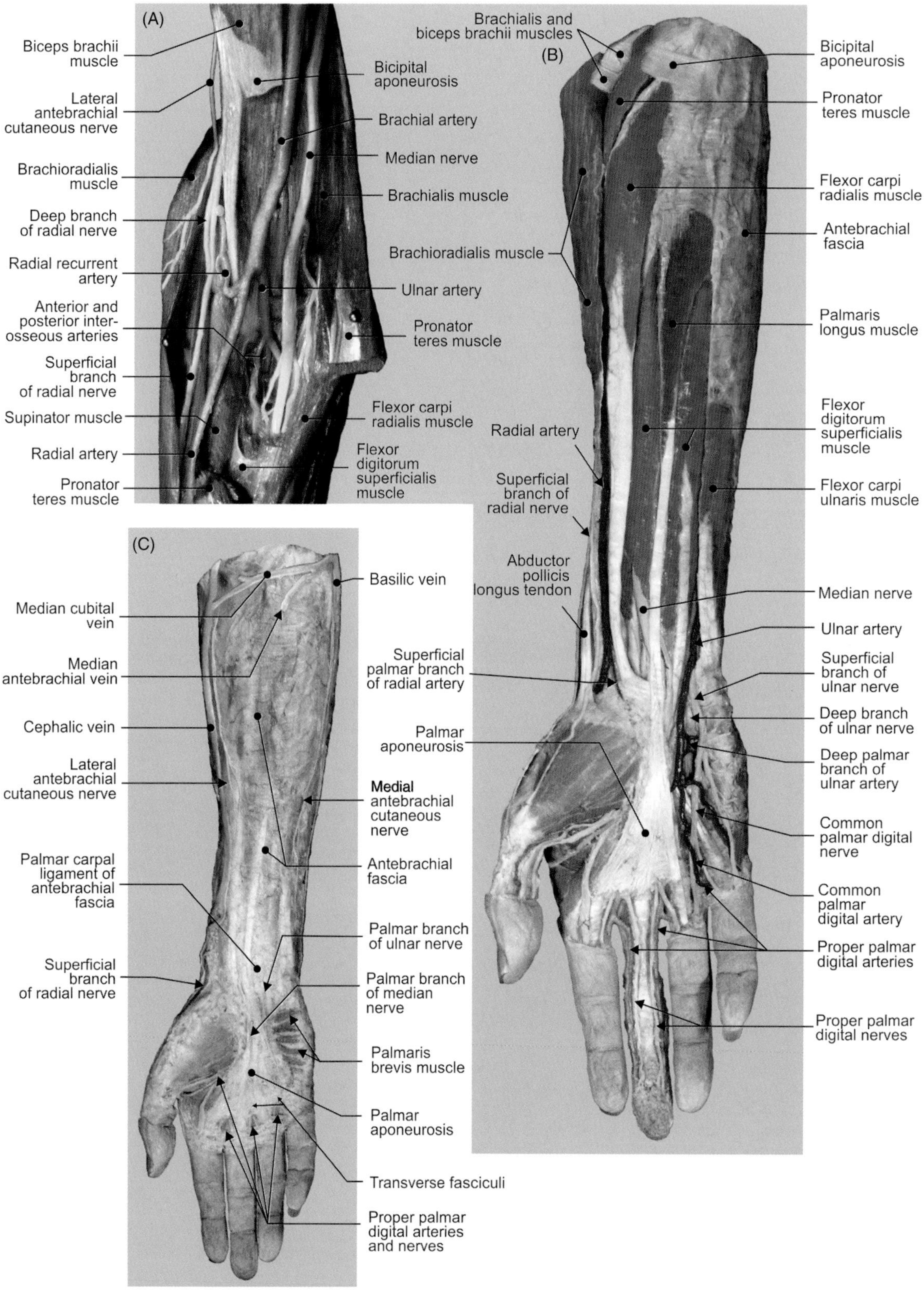

Biceps brachii muscle

Lateral antebrachial cutaneous nerve

Brachioradialis muscle

Deep branch of radial nerve

Radial recurrent artery

Anterior and posterior interosseous arteries

Superficial branch of radial nerve

Supinator muscle

Radial artery

Pronator teres muscle

(A)

Bicipital aponeurosis

Brachial artery

Median nerve

Brachialis muscle

Brachioradialis muscle

Ulnar artery

Pronator teres muscle

Flexor carpi radialis muscle

Flexor digitorum superficialis muscle

(B)

Brachialis and biceps brachii muscles

Bicipital aponeurosis

Pronator teres muscle

Flexor carpi radialis muscle

Antebrachial fascia

Palmaris longus muscle

Flexor digitorum superficialis muscle

Flexor carpi ulnaris muscle

Radial artery

Superficial branch of radial nerve

Abductor pollicis longus tendon

Median nerve

Ulnar artery

Superficial branch of ulnar nerve

Deep branch of ulnar nerve

Deep palmar branch of ulnar artery

Common palmar digital nerve

Common palmar digital artery

Proper palmar digital arteries

Proper palmar digital nerves

(C)

Median cubital vein

Median antebrachial vein

Cephalic vein

Lateral antebrachial cutaneous nerve

Palmar carpal ligament of antebrachial fascia

Superficial branch of radial nerve

Basilic vein

Superficial palmar branch of radial artery

Palmar aponeurosis

Medial antebrachial cutaneous nerve

Antebrachial fascia

Palmar branch of ulnar nerve

Palmar branch of median nerve

Palmaris brevis muscle

Palmar aponeurosis

Transverse fasciculi

Proper palmar digital arteries and nerves

FIGURE 1.12 (A) Anterior cubital region. (B–C) Superficial layer of the anterior antebrachial region.

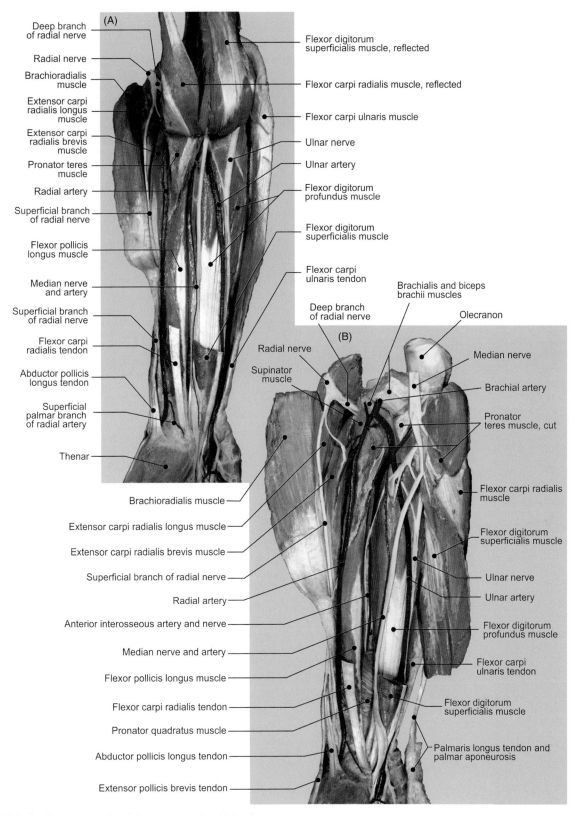

Deep branch of radial nerve

Radial nerve

Brachioradialis muscle

Extensor carpi radialis longus muscle

Extensor carpi radialis brevis muscle

Pronator teres muscle

Radial artery

Superficial branch of radial nerve

Flexor pollicis longus muscle

Median nerve and artery

Superficial branch of radial nerve

Flexor carpi radialis tendon

Abductor pollicis longus tendon

Superficial palmar branch of radial artery

Thenar

(A)

Flexor digitorum superficialis muscle, reflected

Flexor carpi radialis muscle, reflected

Flexor carpi ulnaris muscle

Ulnar nerve

Ulnar artery

Flexor digitorum profundus muscle

Flexor digitorum superficialis muscle

Flexor carpi ulnaris tendon

Deep branch of radial nerve

Radial nerve

Supinator muscle

(B)

Brachialis and biceps brachii muscles

Olecranon

Median nerve

Brachial artery

Pronator teres muscle, cut

Flexor carpi radialis muscle

Flexor digitorum superficialis muscle

Ulnar nerve

Ulnar artery

Flexor digitorum profundus muscle

Flexor carpi ulnaris tendon

Flexor digitorum superficialis muscle

Palmaris longus tendon and palmar aponeurosis

Brachioradialis muscle

Extensor carpi radialis longus muscle

Extensor carpi radialis brevis muscle

Superficial branch of radial nerve

Radial artery

Anterior interosseous artery and nerve

Median nerve and artery

Flexor pollicis longus muscle

Flexor carpi radialis tendon

Pronator quadratus muscle

Abductor pollicis longus tendon

Extensor pollicis brevis tendon

FIGURE 1.13   Deep layer (A–B) of the anterior antebrachial region.

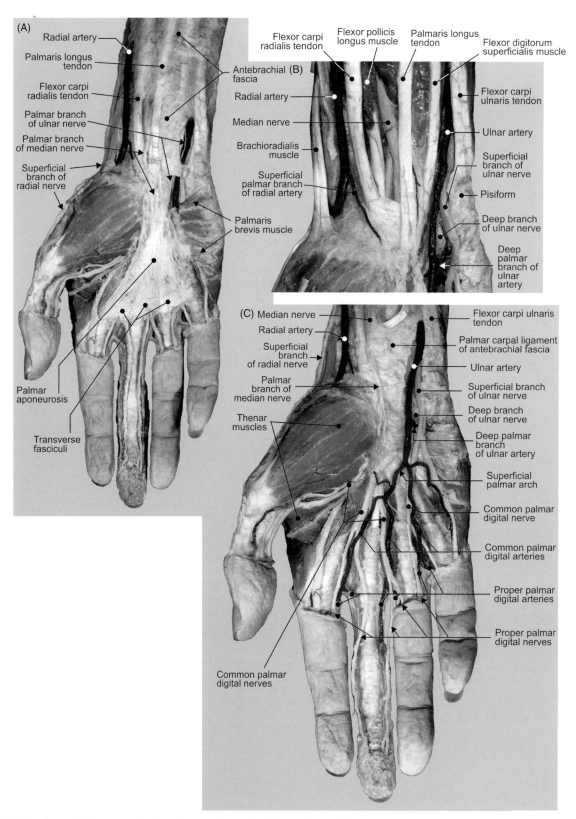

FIGURE 1.14 Superficial structures in the palmar region. (A–C) Different layers offering structural details.

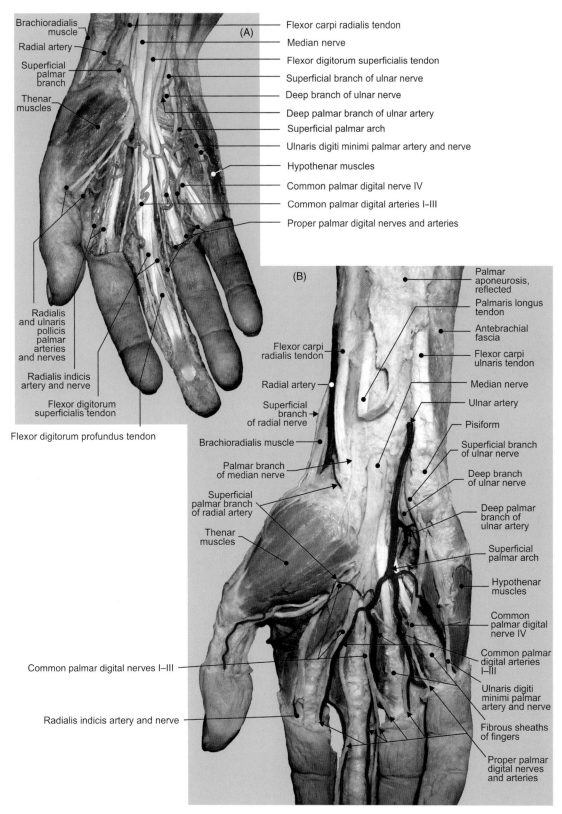

Brachioradialis muscle
Radial artery
Superficial palmar branch
Thenar muscles

(A)
Flexor carpi radialis tendon
Median nerve
Flexor digitorum superficialis tendon
Superficial branch of ulnar nerve
Deep branch of ulnar nerve
Deep palmar branch of ulnar artery
Superficial palmar arch
Ulnaris digiti minimi palmar artery and nerve
Hypothenar muscles
Common palmar digital nerve IV
Common palmar digital arteries I–III
Proper palmar digital nerves and arteries

Radialis and ulnaris pollicis palmar arteries and nerves
Radialis indicis artery and nerve
Flexor digitorum superficialis tendon
Flexor digitorum profundus tendon

(B)
Palmar aponeurosis, reflected
Palmaris longus tendon
Antebrachial fascia
Flexor carpi ulnaris tendon
Median nerve
Ulnar artery
Pisiform
Superficial branch of ulnar nerve
Deep branch of ulnar nerve
Deep palmar branch of ulnar artery
Superficial palmar arch
Hypothenar muscles
Common palmar digital nerve IV
Common palmar digital arteries I–III
Ulnaris digiti minimi palmar artery and nerve
Fibrous sheaths of fingers
Proper palmar digital nerves and arteries

Flexor carpi radialis tendon
Radial artery
Superficial branch of radial nerve
Brachioradialis muscle
Palmar branch of median nerve
Superficial palmar branch of radial artery
Thenar muscles
Common palmar digital nerves I–III
Radialis indicis artery and nerve

FIGURE 1.15 **Structures in the palmar region deep to palmar aponeurosis.** (A–B) Different specimens offering structural details.

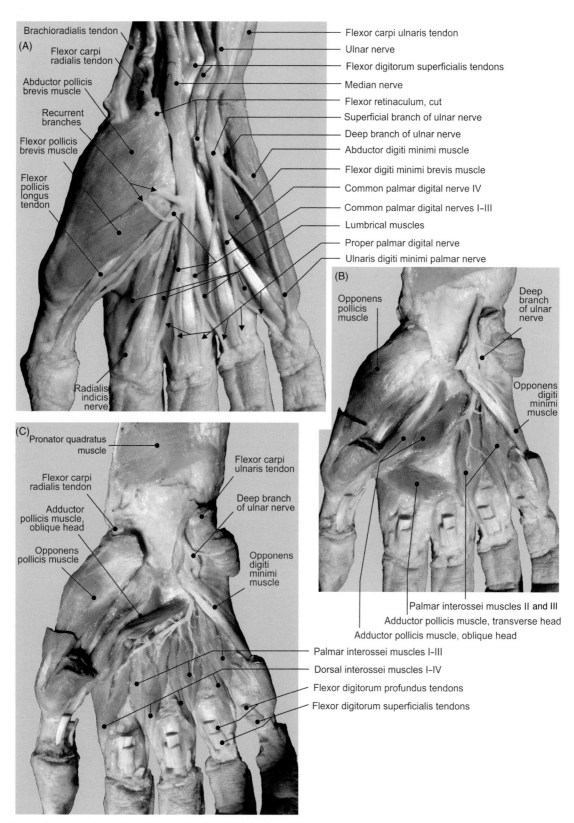

FIGURE 1.16 **Structures in the deep layer of the palmar region.** (A–C) Different layers offering structural details.

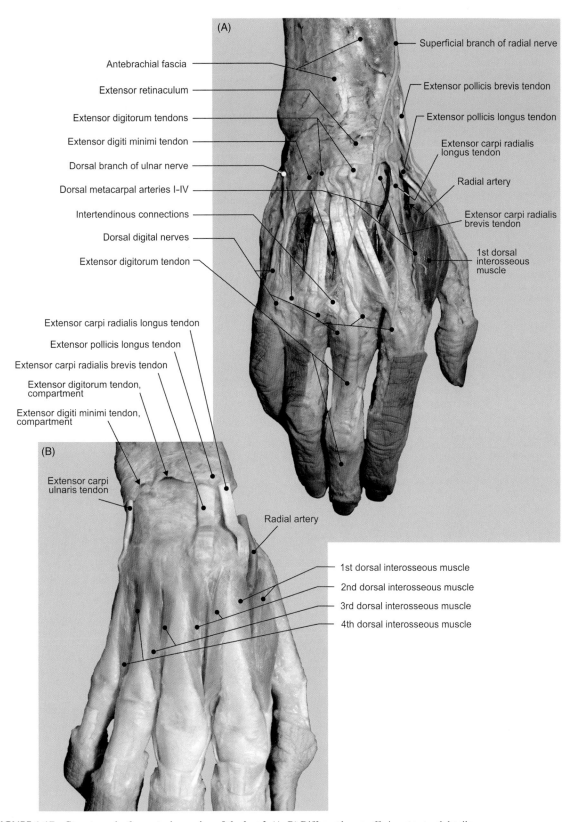

FIGURE 1.17   Structures in the posterior region of the hand. (A–B) Different layers offering structural details.

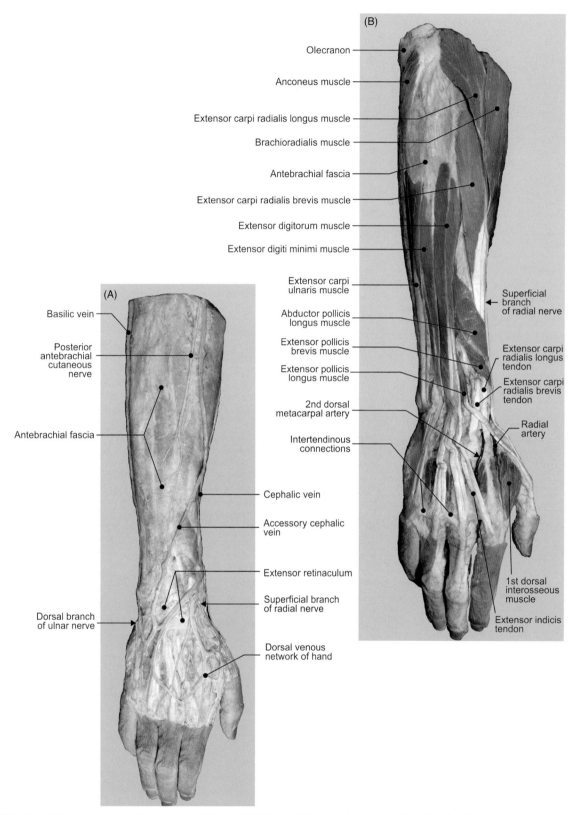

FIGURE 1.18 (A) Structures in the subcutaneous and (B) superficial layers of the posterior antebrachial and hand regions.

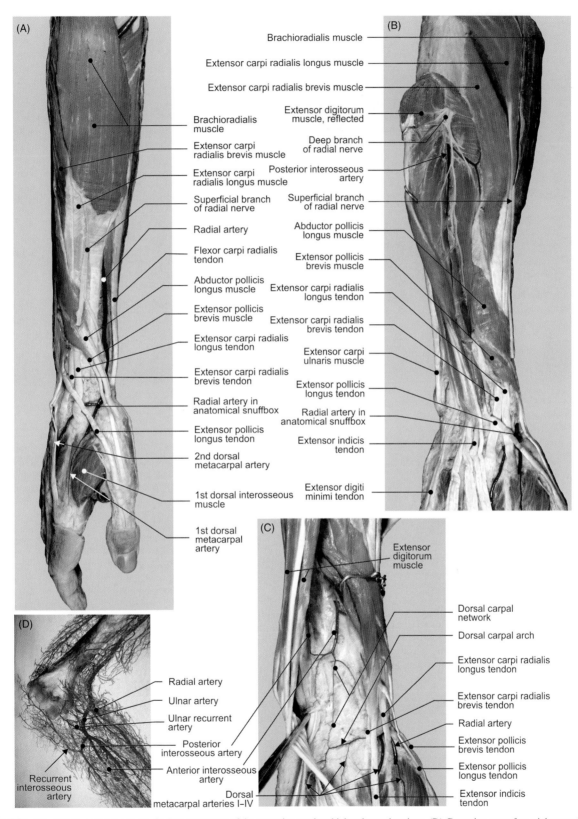

(A)

Brachioradialis muscle

Extensor carpi radialis brevis muscle

Extensor carpi radialis longus muscle

Superficial branch of radial nerve

Radial artery

Flexor carpi radialis tendon

Abductor pollicis longus muscle

Extensor pollicis brevis muscle

Extensor carpi radialis longus tendon

Extensor carpi radialis brevis tendon

Radial artery in anatomical snuffbox

Extensor pollicis longus tendon

2nd dorsal metacarpal artery

1st dorsal interosseous muscle

1st dorsal metacarpal artery

(B)

Brachioradialis muscle

Extensor carpi radialis longus muscle

Extensor carpi radialis brevis muscle

Extensor digitorum muscle, reflected

Deep branch of radial nerve

Posterior interosseous artery

Superficial branch of radial nerve

Abductor pollicis longus muscle

Extensor pollicis brevis muscle

Extensor carpi radialis longus tendon

Extensor carpi radialis brevis tendon

Extensor carpi ulnaris muscle

Extensor pollicis longus tendon

Radial artery in anatomical snuffbox

Extensor indicis tendon

Extensor digiti minimi tendon

(C)

Extensor digitorum muscle

Dorsal carpal network

Dorsal carpal arch

Extensor carpi radialis longus tendon

Extensor carpi radialis brevis tendon

Radial artery

Extensor pollicis brevis tendon

Extensor pollicis longus tendon

Extensor indicis tendon

(D)

Radial artery

Ulnar artery

Ulnar recurrent artery

Posterior interosseous artery

Anterior interosseous artery

Recurrent interosseous artery

Dorsal metacarpal arteries I–IV

FIGURE 1.19   (A) Superficial and (B–C) deep structures of the posterior antebrachial and carpal regions. (D) Corrosion cast of arterial network around the elbow joint.

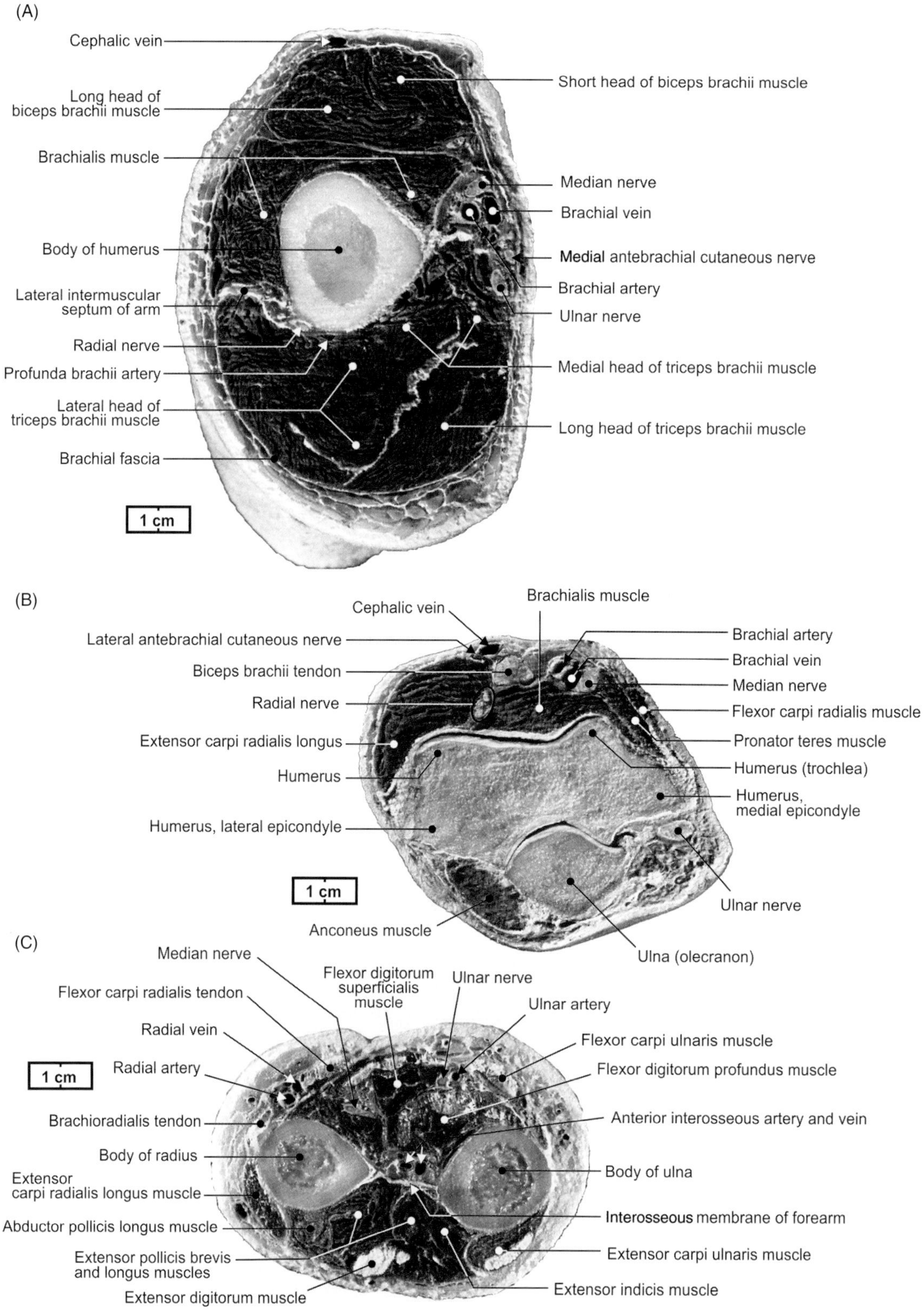

(A)

Cephalic vein

Long head of
biceps brachii muscle

Brachialis muscle

Body of humerus

Lateral intermuscular
septum of arm

Radial nerve

Profunda brachii artery

Lateral head of
triceps brachii muscle

Brachial fascia

Short head of biceps brachii muscle

Median nerve

Brachial vein

Medial antebrachial cutaneous nerve

Brachial artery

Ulnar nerve

Medial head of triceps brachii muscle

Long head of triceps brachii muscle

1 cm

(B)

Brachialis muscle

Cephalic vein

Lateral antebrachial cutaneous nerve

Biceps brachii tendon

Radial nerve

Extensor carpi radialis longus

Humerus

Humerus, lateral epicondyle

Brachial artery

Brachial vein

Median nerve

Flexor carpi radialis muscle

Pronator teres muscle

Humerus (trochlea)

Humerus,
medial epicondyle

Ulnar nerve

1 cm

Anconeus muscle

Ulna (olecranon)

(C)

Median nerve

Flexor carpi radialis tendon

Radial vein

Radial artery

Brachioradialis tendon

Body of radius

Extensor
carpi radialis longus muscle

Abductor pollicis longus muscle

Extensor pollicis brevis
and longus muscles

Extensor digitorum muscle

Flexor digitorum
superficialis
muscle

Ulnar nerve

Ulnar artery

Flexor carpi ulnaris muscle

Flexor digitorum profundus muscle

Anterior interosseous artery and vein

Body of ulna

Interosseous membrane of forearm

Extensor carpi ulnaris muscle

Extensor indicis muscle

1 cm

FIGURE 1.20 Transverse cross-sectional views of the right (A) arm, (B) elbow joint, and (C) forearm.

(A)

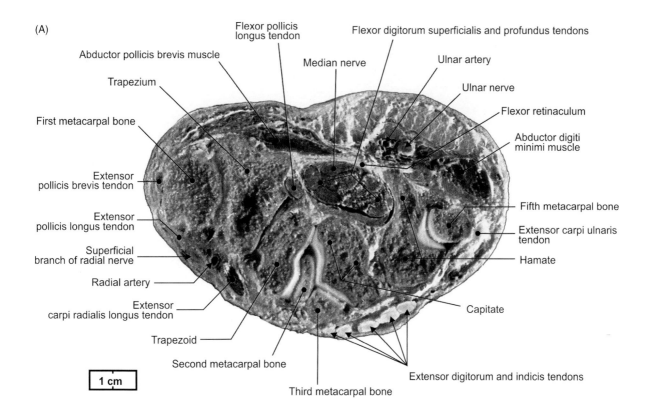

Flexor pollicis longus tendon

Flexor digitorum superficialis and profundus tendons

Abductor pollicis brevis muscle

Median nerve

Ulnar artery

Trapezium

Ulnar nerve

First metacarpal bone

Flexor retinaculum

Extensor pollicis brevis tendon

Abductor digiti minimi muscle

Extensor pollicis longus tendon

Fifth metacarpal bone

Superficial branch of radial nerve

Extensor carpi ulnaris tendon

Radial artery

Hamate

Extensor carpi radialis longus tendon

Capitate

Trapezoid

Extensor digitorum and indicis tendons

Second metacarpal bone

Third metacarpal bone

1 cm

(B)

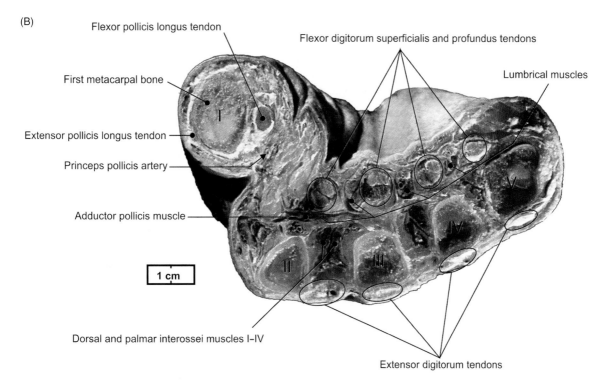

Flexor pollicis longus tendon

Flexor digitorum superficialis and profundus tendons

First metacarpal bone

Lumbrical muscles

Extensor pollicis longus tendon

Princeps pollicis artery

Adductor pollicis muscle

1 cm

Dorsal and palmar interossei muscles I–IV

Extensor digitorum tendons

FIGURE 1.21 Transverse cross-sectional views of the right (A) carpal and (B) metacarpal regions.

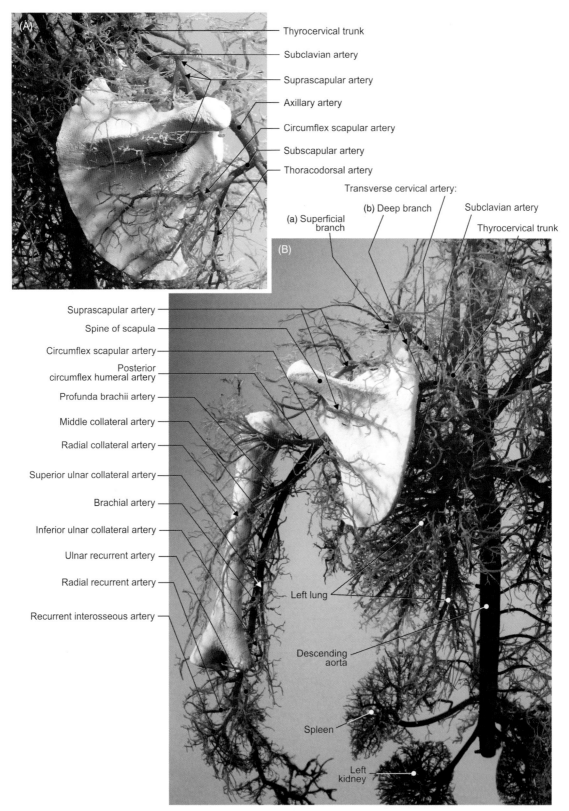

FIGURE 1.22   Fetal arterial corrosion casts on the posterior aspects of (A) scapula and (B) scapula, brachium, and elbow.

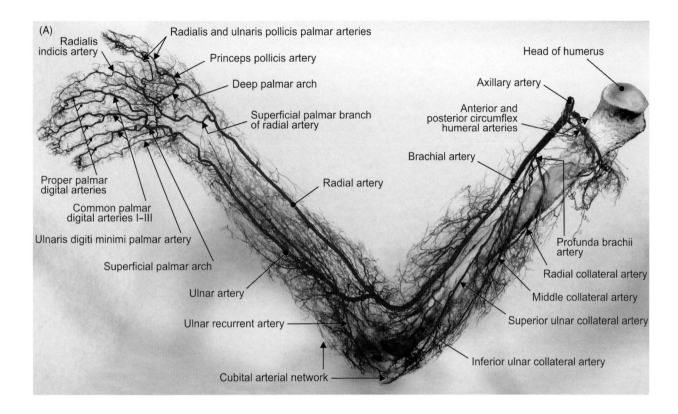

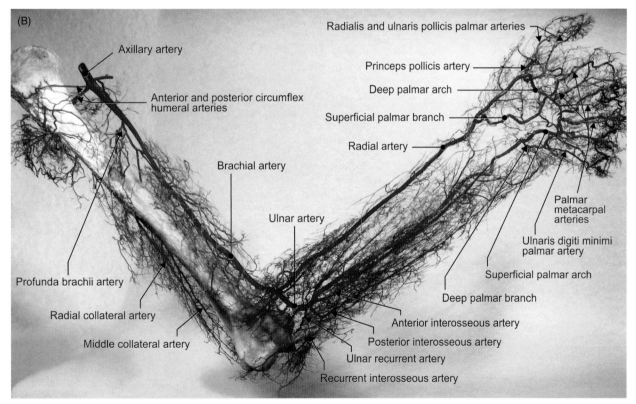

FIGURE 1.23 **Arterial corrosion cast of the right upper limb.** (A) Medial and (B) lateral views.

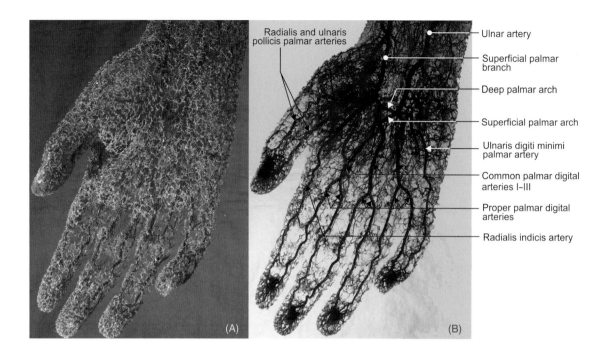

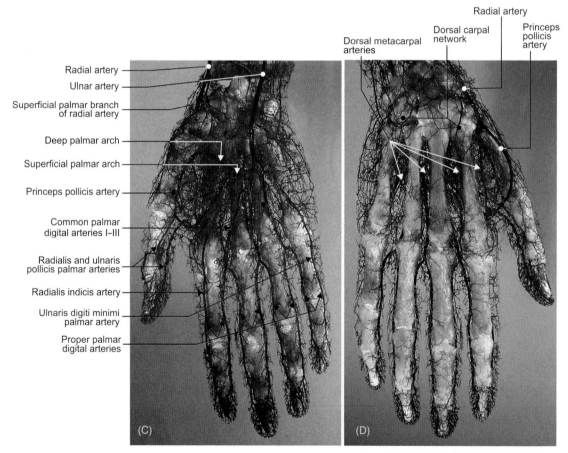

FIGURE 1.24   Arterial networks (corrosion casts) of right hand. (A) Superficial vessels, (B) major vessels, (C) palmar side, and (D) dorsal side.

Chapter 2

Lower Limb and Vascularization

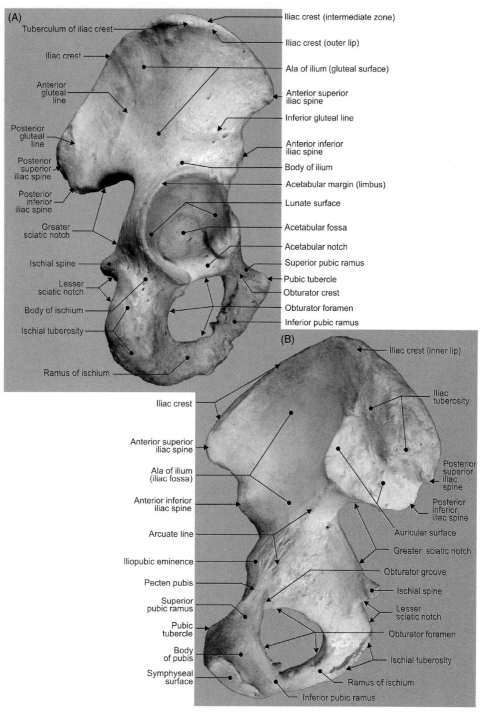

(A)
- Tuberculum of iliac crest
- Iliac crest
- Anterior gluteal line
- Posterior gluteal line
- Posterior superior iliac spine
- Posterior inferior iliac spine
- Greater sciatic notch
- Ischial spine
- Lesser sciatic notch
- Body of ischium
- Ischial tuberosity
- Ramus of ischium

- Iliac crest (intermediate zone)
- Iliac crest (outer lip)
- Ala of ilium (gluteal surface)
- Anterior superior iliac spine
- Inferior gluteal line
- Anterior inferior iliac spine
- Body of ilium
- Acetabular margin (limbus)
- Lunate surface
- Acetabular fossa
- Acetabular notch
- Superior pubic ramus
- Pubic tubercle
- Obturator crest
- Obturator foramen
- Inferior pubic ramus

(B)
- Iliac crest
- Anterior superior iliac spine
- Ala of ilium (iliac fossa)
- Anterior inferior iliac spine
- Arcuate line
- Iliopubic eminence
- Pecten pubis
- Superior pubic ramus
- Pubic tubercle
- Body of pubis
- Symphyseal surface

- Iliac crest (inner lip)
- Iliac tuberosity
- Posterior superior iliac spine
- Posterior inferior iliac spine
- Auricular surface
- Greater sciatic notch
- Obturator groove
- Ischial spine
- Lesser sciatic notch
- Obturator foramen
- Ischial tuberosity
- Ramus of ischium
- Inferior pubic ramus

FIGURE 2.1   **The hip bone.** (A) Lateral and (B) medial views.

Atlas of the Human Body
Copyright © 2017 Elsevier Inc. All rights reserved.

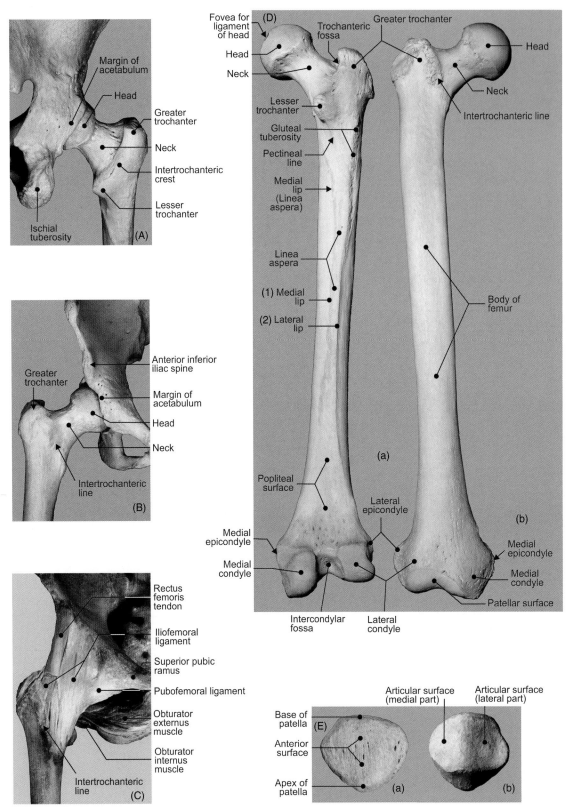

FIGURE 2.2 The hip joint: (A) posterior and anterior (B) views of the bones; (C) anterior view of the ligaments. (D) Right femur: (a) posterior and (b) anterior views. (E) Patella: (a) anterior and (b) posterior views.

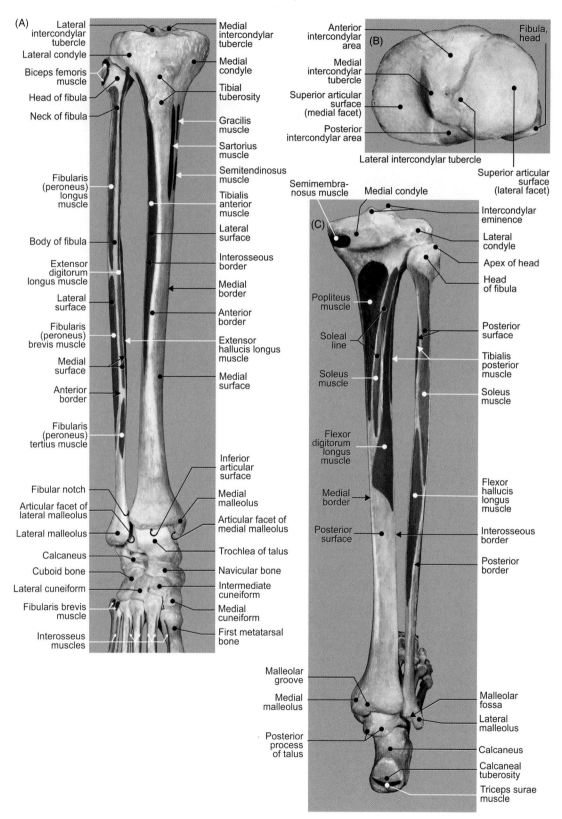

FIGURE 2.3 **Right tibia and fibula.** (A) Anterior, (B) superior, and (C) posterior views.

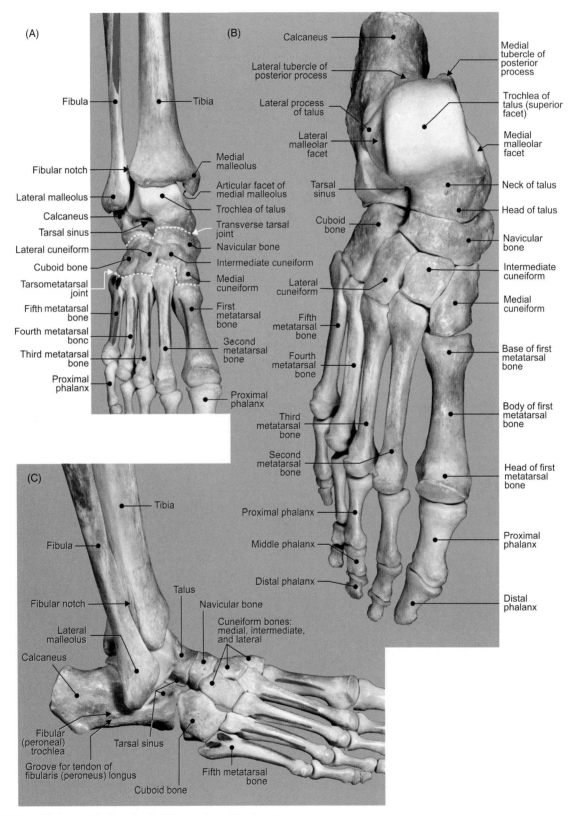

(A)

Fibula
Tibia

Fibular notch
Lateral malleolus
Calcaneus
Tarsal sinus
Lateral cuneiform
Cuboid bone
Tarsometatarsal joint
Fifth metatarsal bone
Fourth metatarsal bone
Third metatarsal bone
Proximal phalanx

Medial malleolus
Articular facet of medial malleolus
Trochlea of talus
Transverse tarsal joint
Navicular bone
Intermediate cuneiform
Medial cuneiform
First metatarsal bone
Second metatarsal bone
Proximal phalanx

(B)

Calcaneus
Lateral tubercle of posterior process
Lateral process of talus
Lateral malleolar facet
Tarsal sinus
Cuboid bone
Lateral cuneiform
Fifth metatarsal bone
Fourth metatarsal bone
Third metatarsal bone
Second metatarsal bone
Proximal phalanx
Middle phalanx
Distal phalanx

Medial tubercle of posterior process
Trochlea of talus (superior facet)
Medial malleolar facet
Neck of talus
Head of talus
Navicular bone
Intermediate cuneiform
Medial cuneiform
Base of first metatarsal bone
Body of first metatarsal bone
Head of first metatarsal bone
Proximal phalanx
Distal phalanx

(C)

Tibia
Fibula
Fibular notch
Lateral malleolus
Calcaneus
Fibular (peroneal) trochlea
Groove for tendon of fibularis (peroneus) longus
Tarsal sinus
Cuboid bone

Talus
Navicular bone
Cuneiform bones: medial, intermediate, and lateral
Fifth metatarsal bone

FIGURE 2.4 Skeleton of the foot. (A–C) Different views of the dorsal aspect.

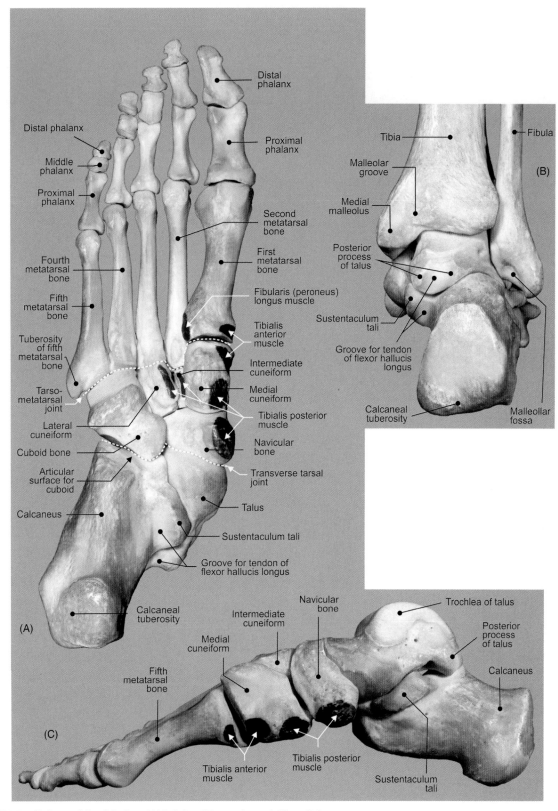

FIGURE 2.5   Skeleton of the right foot. (A) Inferior, (B) posterior, and (C) medial aspects.

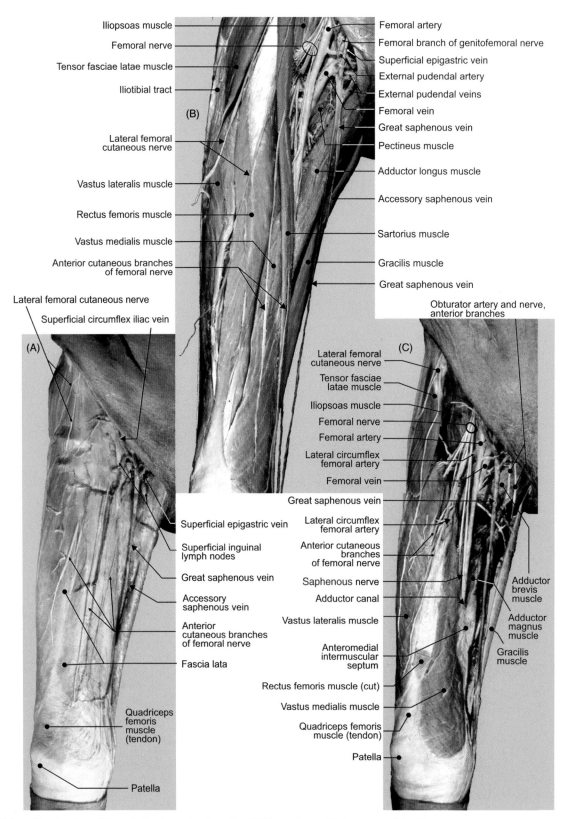

Iliopsoas muscle

Femoral nerve

Tensor fasciae latae muscle

Iliotibial tract

(B)

Lateral femoral
cutaneous nerve

Vastus lateralis muscle

Rectus femoris muscle

Vastus medialis muscle

Anterior cutaneous branches
of femoral nerve

Lateral femoral cutaneous nerve

Superficial circumflex iliac vein

(A)

Femoral artery

Femoral branch of genitofemoral nerve

Superficial epigastric vein

External pudendal artery

External pudendal veins

Femoral vein

Great saphenous vein

Pectineus muscle

Adductor longus muscle

Accessory saphenous vein

Sartorius muscle

Gracilis muscle

Great saphenous vein

Obturator artery and nerve,
anterior branches

(C)

Lateral femoral
cutaneous nerve

Tensor fasciae
latae muscle

Iliopsoas muscle

Femoral nerve

Femoral artery

Lateral circumflex
femoral artery

Femoral vein

Great saphenous vein

Lateral circumflex
femoral artery

Anterior cutaneous
branches
of femoral nerve

Saphenous nerve

Adductor canal

Vastus lateralis muscle

Anteromedial
intermuscular
septum

Rectus femoris muscle (cut)

Vastus medialis muscle

Quadriceps femoris
muscle (tendon)

Patella

Adductor
brevis
muscle

Adductor
magnus
muscle

Gracilis
muscle

Superficial epigastric vein

Superficial inguinal
lymph nodes

Great saphenous vein

Accessory
saphenous vein

Anterior
cutaneous branches
of femoral nerve

Fascia lata

Quadriceps
femoris
muscle
(tendon)

Patella

FIGURE 2.6 **Frontal view of the anterior femoral region.** (A–C) Different layers offering structural details.

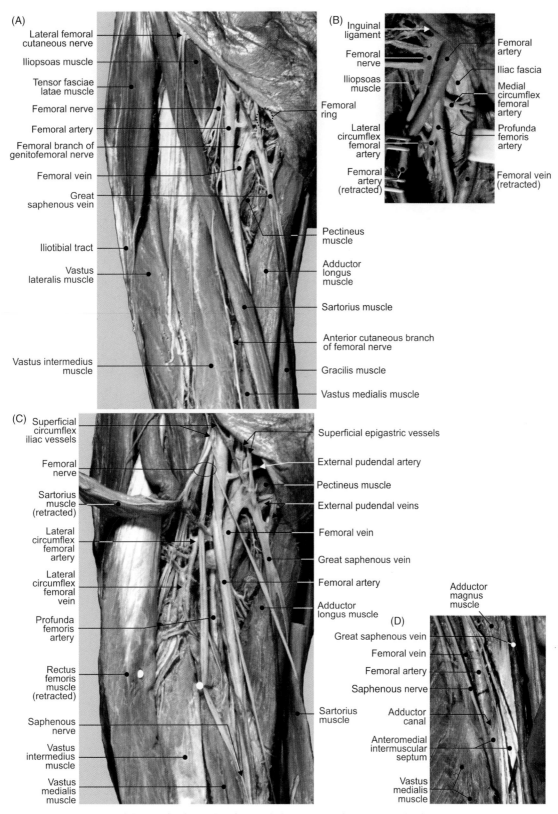

FIGURE 2.7 **Frontal view of the right anterior femoral region.** (A,C) Frontal view of the right anterior femoral region, (B) femoral artery, and (D) detail of the lower medial femoral subregion.

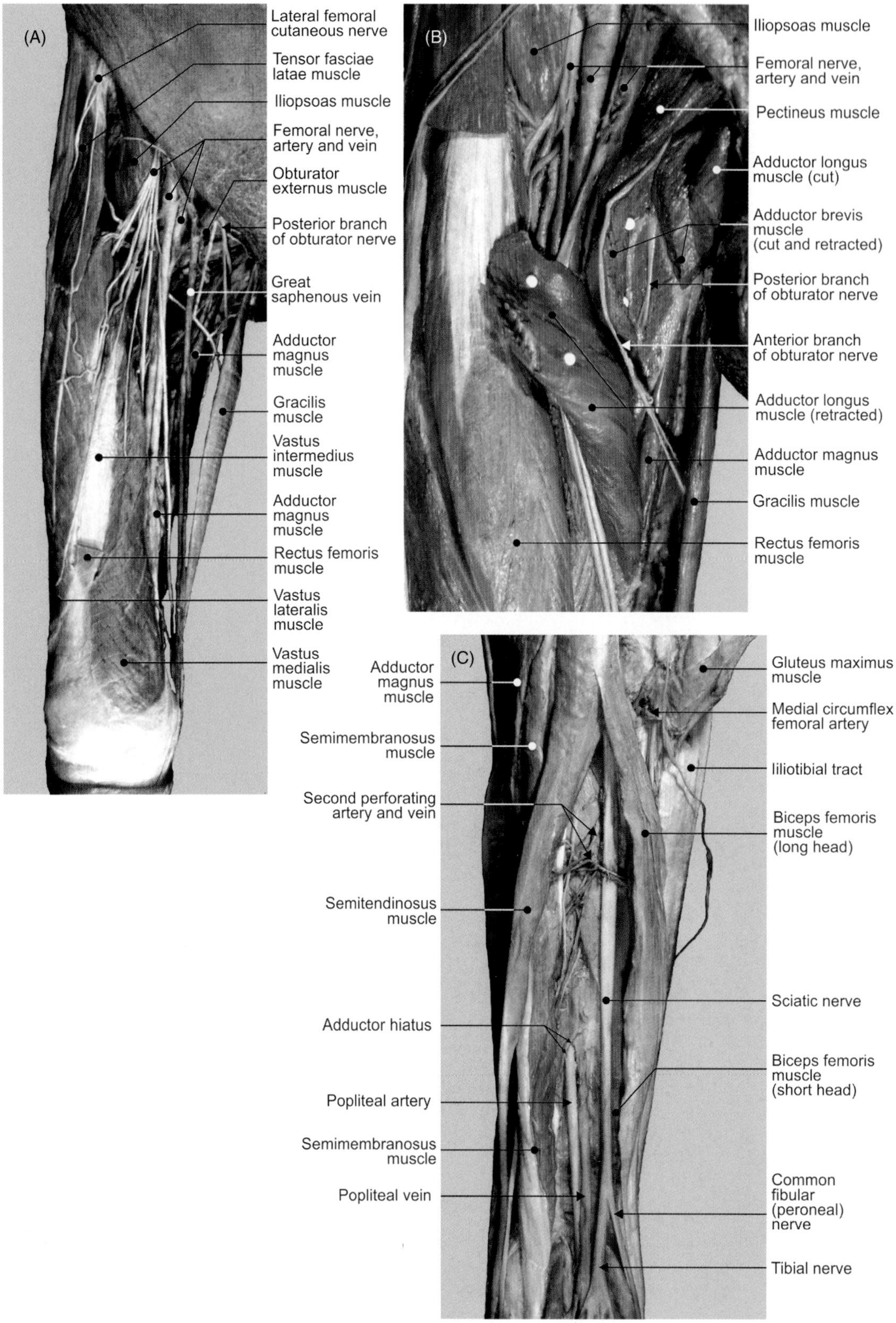

(A)

Lateral femoral cutaneous nerve

Tensor fasciae latae muscle

Iliopsoas muscle

Femoral nerve, artery and vein

Obturator externus muscle

Posterior branch of obturator nerve

Great saphenous vein

Adductor magnus muscle

Gracilis muscle

Vastus intermedius muscle

Adductor magnus muscle

Rectus femoris muscle

Vastus lateralis muscle

Vastus medialis muscle

(B)

Iliopsoas muscle

Femoral nerve, artery and vein

Pectineus muscle

Adductor longus muscle (cut)

Adductor brevis muscle (cut and retracted)

Posterior branch of obturator nerve

Anterior branch of obturator nerve

Adductor longus muscle (retracted)

Adductor magnus muscle

Gracilis muscle

Rectus femoris muscle

(C)

Adductor magnus muscle

Semimembranosus muscle

Second perforating artery and vein

Semitendinosus muscle

Adductor hiatus

Popliteal artery

Semimembranosus muscle

Popliteal vein

Gluteus maximus muscle

Medial circumflex femoral artery

Iliotibial tract

Biceps femoris muscle (long head)

Sciatic nerve

Biceps femoris muscle (short head)

Common fibular (peroneal) nerve

Tibial nerve

FIGURE 2.8   The right lower limb. Frontal views of the (A–B) anterior femoral regions and (C) posterior femoral region.

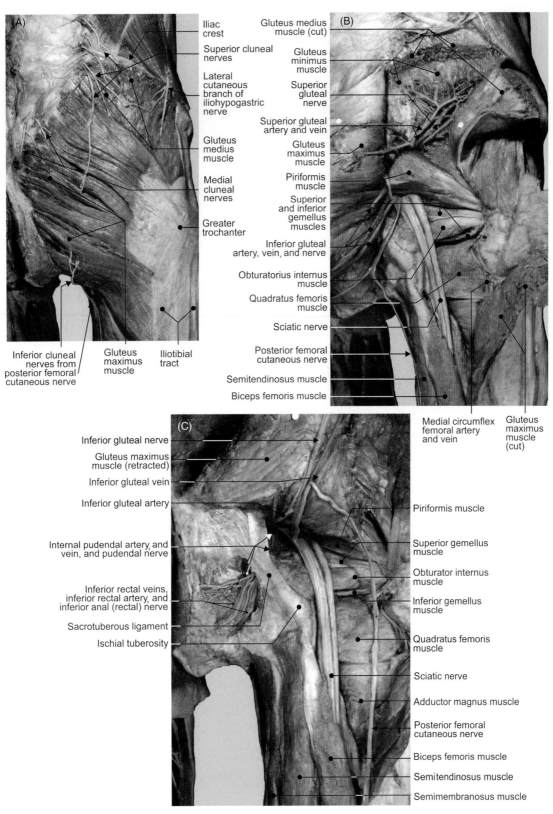

FIGURE 2.9   **The right gluteal region.** (A–C) Different views offering structural details of the superficial and deeper layers.

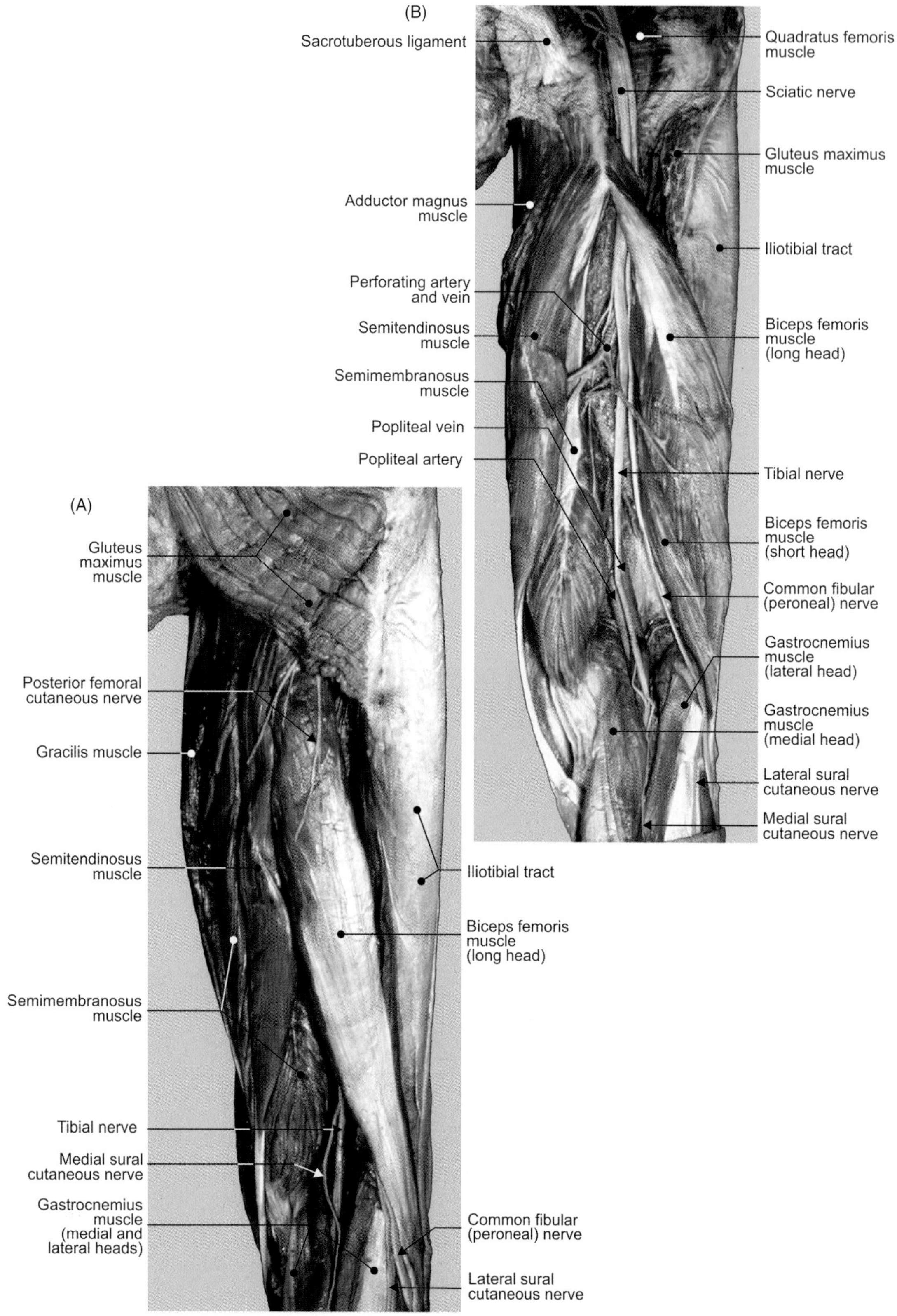

(B)

Sacrotuberous ligament

Adductor magnus
muscle

Perforating artery
and vein

Semitendinosus
muscle

Semimembranosus
muscle

Popliteal vein

Popliteal artery

Quadratus femoris
muscle

Sciatic nerve

Gluteus maximus
muscle

Iliotibial tract

Biceps femoris
muscle
(long head)

Tibial nerve

Biceps femoris
muscle
(short head)

Common fibular
(peroneal) nerve

Gastrocnemius
muscle
(lateral head)

Gastrocnemius
muscle
(medial head)

Lateral sural
cutaneous nerve

Medial sural
cutaneous nerve

(A)

Gluteus
maximus
muscle

Posterior femoral
cutaneous nerve

Gracilis muscle

Semitendinosus
muscle

Semimembranosus
muscle

Tibial nerve

Medial sural
cutaneous nerve

Gastrocnemius
muscle
(medial and
lateral heads)

Iliotibial tract

Biceps femoris
muscle
(long head)

Common fibular
(peroneal) nerve

Lateral sural
cutaneous nerve

FIGURE 2.10 The right posterior femoral regions and popliteal fossa. (A) Superficial and (B) deep layers.

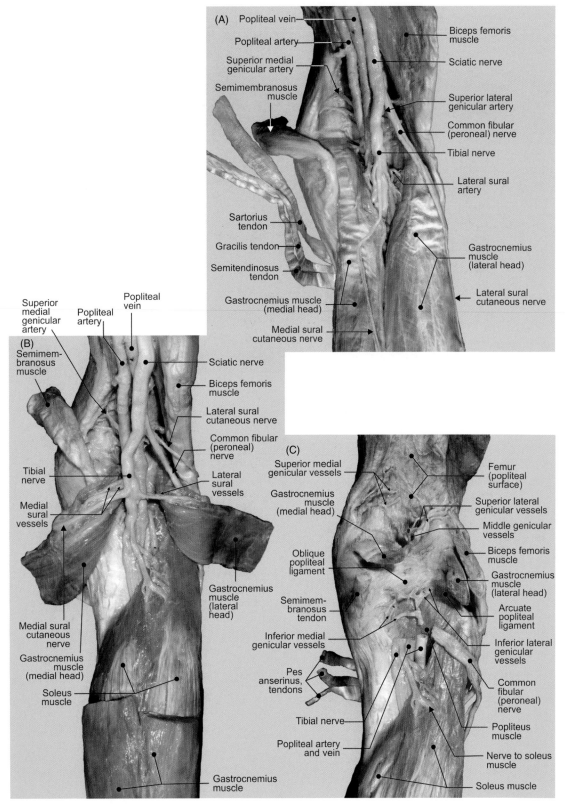

FIGURE 2.11 **The right posterior view of the knee region and popliteal fossa.** (A–C) Different views offering structural details of the walls and content.

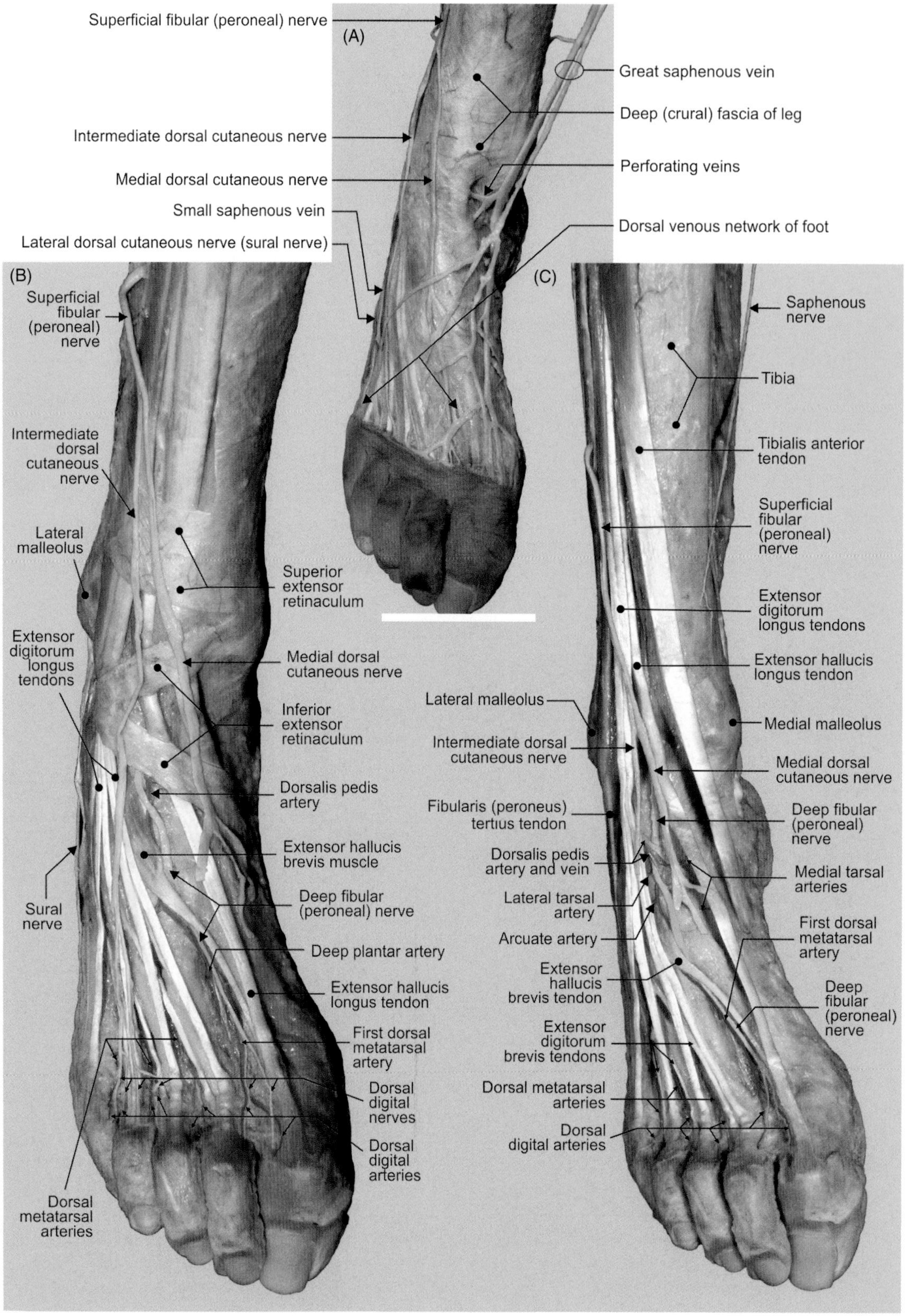

FIGURE 2.12 **Nerves, blood vessels, and tendons of the right foot.** (A) Superficial and (B–C) deep layers.

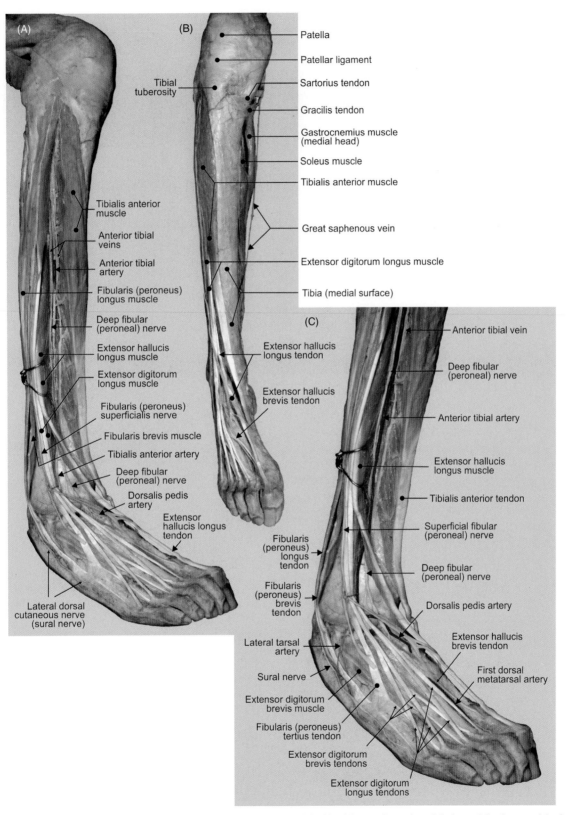

FIGURE 2.13   The right lower limb. (A–C) Different views offering structural details of the anterior region of the leg and the dorsum of the foot.

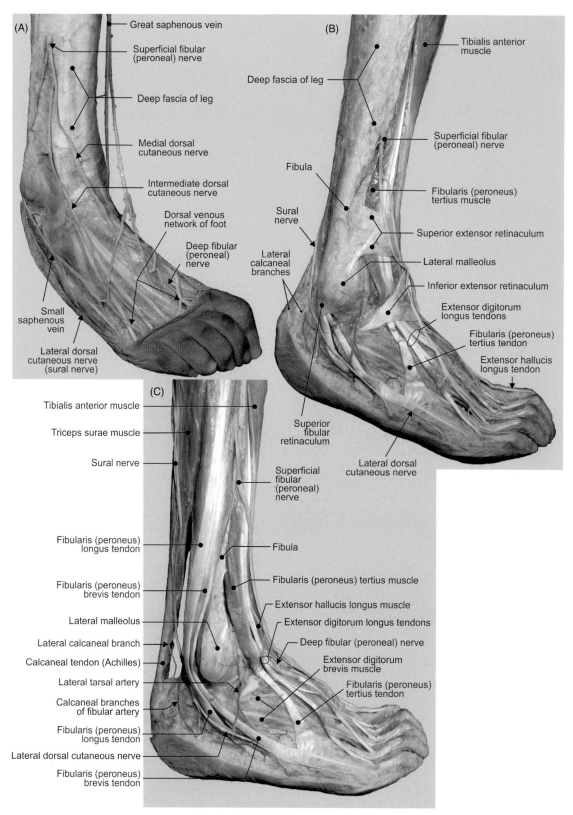

FIGURE 2.14 **The right foot.** (A–C) Different layers offering the lateral aspect of nerves, blood vessels, and muscles on the dorsal surface.

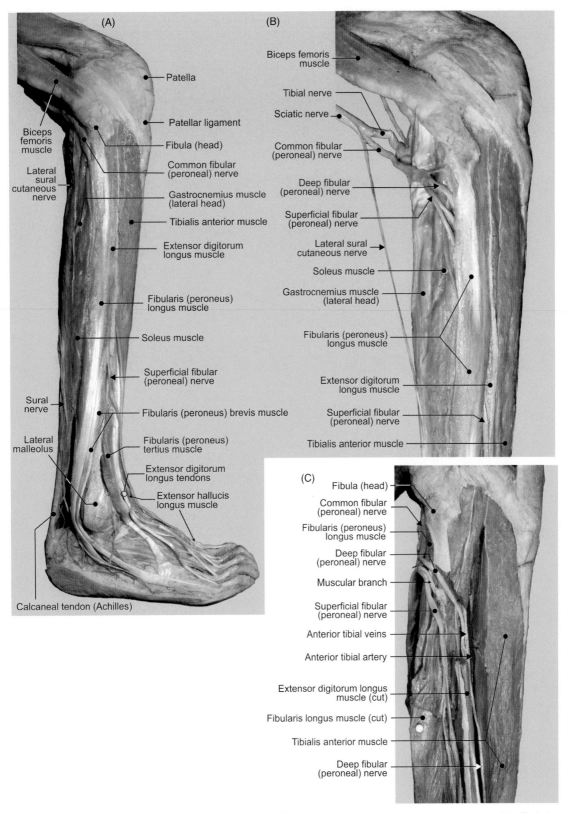

(A)

- Patella
- Patellar ligament
- Fibula (head)
- Common fibular (peroneal) nerve
- Gastrocnemius muscle (lateral head)
- Tibialis anterior muscle
- Extensor digitorum longus muscle
- Fibularis (peroneus) longus muscle
- Soleus muscle
- Superficial fibular (peroneal) nerve
- Fibularis (peroneus) brevis muscle
- Fibularis (peroneus) tertius muscle
- Extensor digitorum longus tendons
- Extensor hallucis longus muscle

Biceps femoris muscle
Lateral sural cutaneous nerve
Sural nerve
Lateral malleolus
Calcaneal tendon (Achilles)

(B)

Biceps femoris muscle
Tibial nerve
Sciatic nerve
Common fibular (peroneal) nerve
Deep fibular (peroneal) nerve
Superficial fibular (peroneal) nerve
Lateral sural cutaneous nerve
Soleus muscle
Gastrocnemius muscle (lateral head)
Fibularis (peroneus) longus muscle
Extensor digitorum longus muscle
Superficial fibular (peroneal) nerve
Tibialis anterior muscle

(C)

- Fibula (head)
- Common fibular (peroneal) nerve
- Fibularis (peroneus) longus muscle
- Deep fibular (peroneal) nerve
- Muscular branch
- Superficial fibular (peroneal) nerve
- Anterior tibial veins
- Anterior tibial artery
- Extensor digitorum longus muscle (cut)
- Fibularis longus muscle (cut)
- Tibialis anterior muscle
- Deep fibular (peroneal) nerve

FIGURE 2.15 **The right limb.** (A) Anterior region of leg with dorsum of foot and (B) without foot. (C) Removed upper part of the fibularis (peroneus) longus muscle.

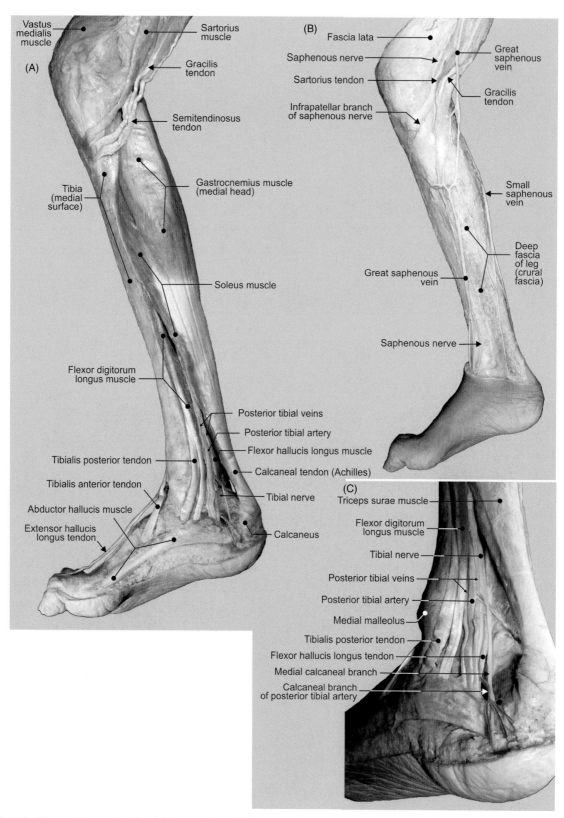

Vastus medialis muscle

Sartorius muscle

(A)

Gracilis tendon

Semitendinosus tendon

Tibia (medial surface)

Gastrocnemius muscle (medial head)

Soleus muscle

Flexor digitorum longus muscle

Posterior tibial veins

Posterior tibial artery

Tibialis posterior tendon

Flexor hallucis longus muscle

Tibialis anterior tendon

Calcaneal tendon (Achilles)

Abductor hallucis muscle

Tibial nerve

Extensor hallucis longus tendon

Calcaneus

(B)

Fascia lata

Great saphenous vein

Saphenous nerve

Sartorius tendon

Gracilis tendon

Infrapatellar branch of saphenous nerve

Small saphenous vein

Deep fascia of leg (crural fascia)

Great saphenous vein

Saphenous nerve

(C)

Triceps surae muscle

Flexor digitorum longus muscle

Tibial nerve

Posterior tibial veins

Posterior tibial artery

Medial malleolus

Tibialis posterior tendon

Flexor hallucis longus tendon

Medial calcaneal branch

Calcaneal branch of posterior tibial artery

FIGURE 2.16 The medial aspects of the right leg and foot. (A) Deep and (B) superficial dissected layers. (C) The medial retromalleolar region.

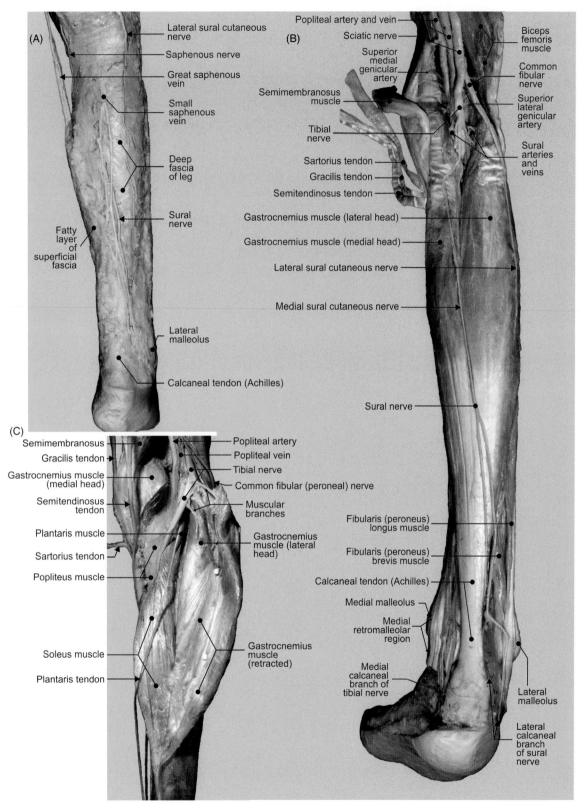

FIGURE 2.17 **The popliteal and posterior crural regions of right limb.** (A–B) Superficial and (C) deep layers.

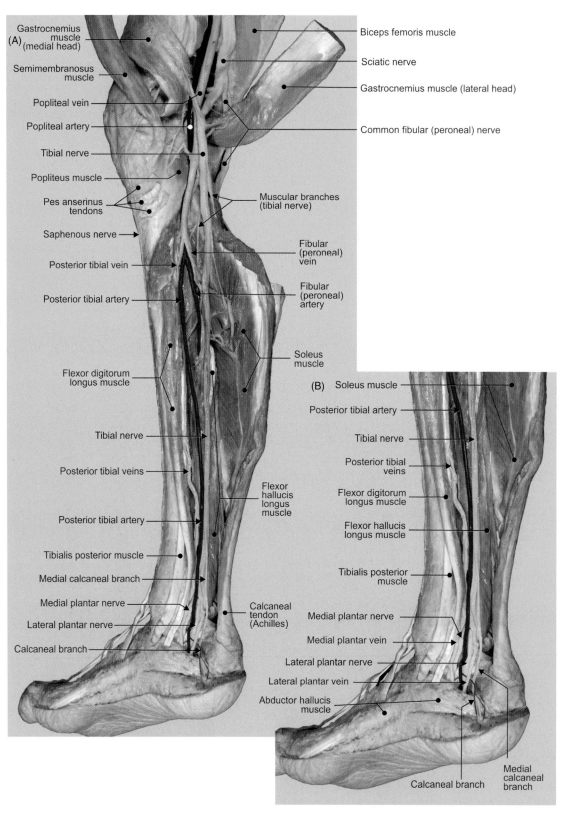

FIGURE 2.18 (A–B) Structural details of the deep layers of the popliteal, posterior crural, and medial retromalleolar regions.

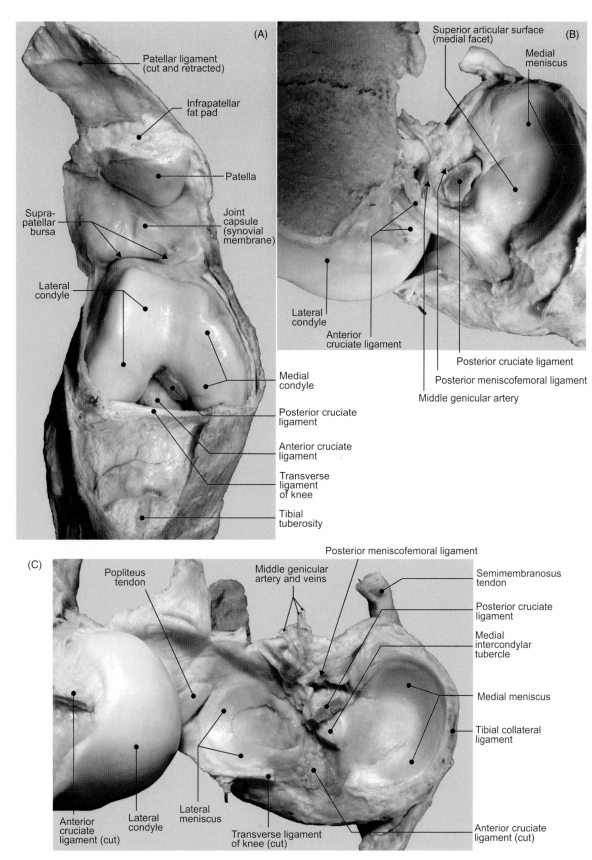

FIGURE 2.19 **The right knee joint.** (A) Anterior view of open joint. (B) Removed medial condyle and transected posterior cruciate ligament. (C) Superior view of the tibial articular surface after transection of both cruciate ligaments.

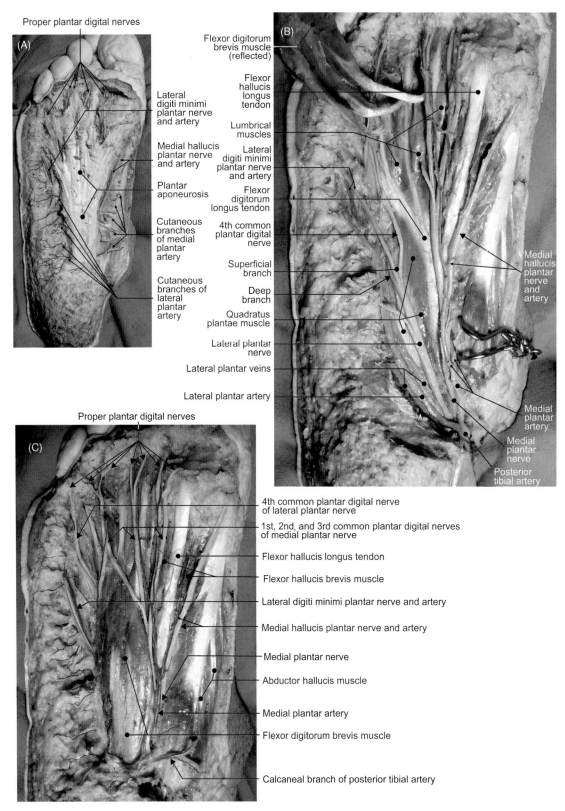

Proper plantar digital nerves

(A)

Lateral digiti minimi plantar nerve and artery

Medial hallucis plantar nerve and artery

Plantar aponeurosis

Cutaneous branches of medial plantar artery

Cutaneous branches of lateral plantar artery

(B)

Flexor digitorum brevis muscle (reflected)

Flexor hallucis longus tendon

Lumbrical muscles

Lateral digiti minimi plantar nerve and artery

Flexor digitorum longus tendon

4th common plantar digital nerve

Superficial branch

Deep branch

Quadratus plantae muscle

Lateral plantar nerve

Lateral plantar veins

Lateral plantar artery

Medial hallucis plantar nerve and artery

Medial plantar artery

Medial plantar nerve

Posterior tibial artery

Proper plantar digital nerves

(C)

4th common plantar digital nerve of lateral plantar nerve

1st, 2nd, and 3rd common plantar digital nerves of medial plantar nerve

Flexor hallucis longus tendon

Flexor hallucis brevis muscle

Lateral digiti minimi plantar nerve and artery

Medial hallucis plantar nerve and artery

Medial plantar nerve

Abductor hallucis muscle

Medial plantar artery

Flexor digitorum brevis muscle

Calcaneal branch of posterior tibial artery

FIGURE 2.20   Plantar aspect of the right foot. Structures in the (A) superficial and (B–C) middle layers.

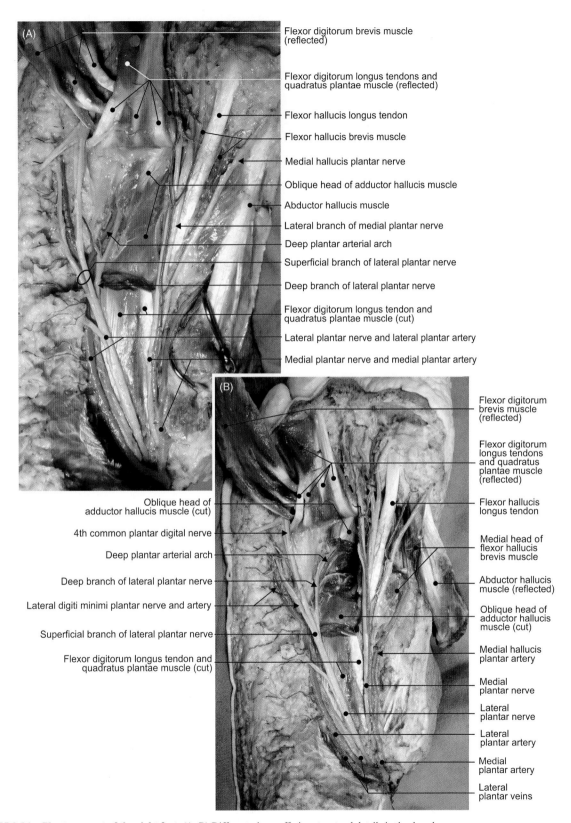

Flexor digitorum brevis muscle (reflected)

Flexor digitorum longus tendons and quadratus plantae muscle (reflected)

Flexor hallucis longus tendon

Flexor hallucis brevis muscle

Medial hallucis plantar nerve

Oblique head of adductor hallucis muscle

Abductor hallucis muscle

Lateral branch of medial plantar nerve

Deep plantar arterial arch

Superficial branch of lateral plantar nerve

Deep branch of lateral plantar nerve

Flexor digitorum longus tendon and quadratus plantae muscle (cut)

Lateral plantar nerve and lateral plantar artery

Medial plantar nerve and medial plantar artery

Oblique head of adductor hallucis muscle (cut)

4th common plantar digital nerve

Deep plantar arterial arch

Deep branch of lateral plantar nerve

Lateral digiti minimi plantar nerve and artery

Superficial branch of lateral plantar nerve

Flexor digitorum longus tendon and quadratus plantae muscle (cut)

Flexor digitorum brevis muscle (reflected)

Flexor digitorum longus tendons and quadratus plantae muscle (reflected)

Flexor hallucis longus tendon

Medial head of flexor hallucis brevis muscle

Abductor hallucis muscle (reflected)

Oblique head of adductor hallucis muscle (cut)

Medial hallucis plantar artery

Medial plantar nerve

Lateral plantar nerve

Lateral plantar artery

Medial plantar artery

Lateral plantar veins

FIGURE 2.21 **Plantar aspect of the right foot.** (A–B) Different views offering structural details in the deep layer.

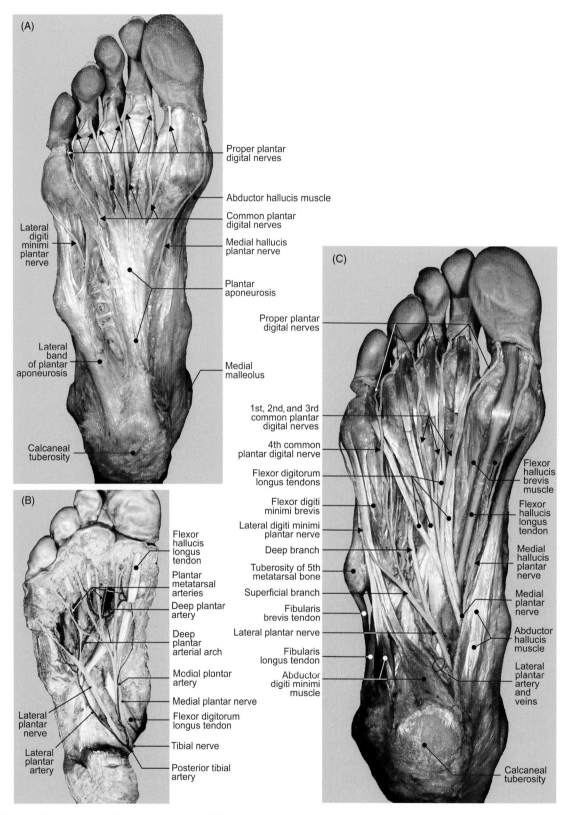

(A)

Proper plantar digital nerves

Abductor hallucis muscle

Common plantar digital nerves

Medial hallucis plantar nerve

Plantar aponeurosis

Lateral digiti minimi plantar nerve

Lateral band of plantar aponeurosis

Medial malleolus

Calcaneal tuberosity

(B)

Flexor hallucis longus tendon

Plantar metatarsal arteries

Deep plantar artery

Deep plantar arterial arch

Medial plantar artery

Medial plantar nerve

Flexor digitorum longus tendon

Tibial nerve

Posterior tibial artery

Lateral plantar nerve

Lateral plantar artery

(C)

Proper plantar digital nerves

1st, 2nd, and 3rd common plantar digital nerves

4th common plantar digital nerve

Flexor digitorum longus tendons

Flexor digiti minimi brevis

Lateral digiti minimi plantar nerve

Deep branch

Tuberosity of 5th metatarsal bone

Superficial branch

Fibularis brevis tendon

Lateral plantar nerve

Fibularis longus tendon

Abductor digiti minimi muscle

Flexor hallucis brevis muscle

Flexor hallucis longus tendon

Medial hallucis plantar nerve

Medial plantar nerve

Abductor hallucis muscle

Lateral plantar artery and veins

Calcaneal tuberosity

FIGURE 2.22 **Plantar region of the right foot.** (A–C) Different views displaying arteries, nerves, and tendons.

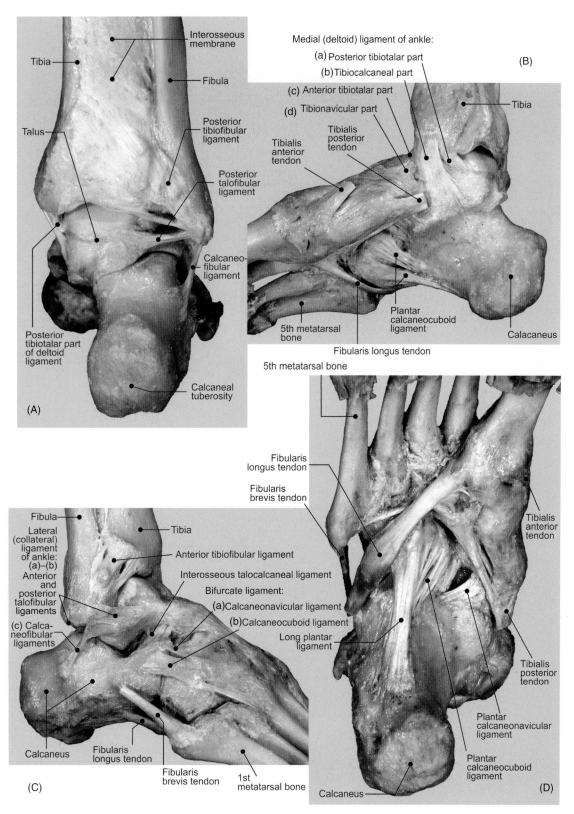

FIGURE 2.23 **Ligaments and tendons of the right foot.** (A) Posterior, (B) medial, (C) lateral, and (D) inferior views.

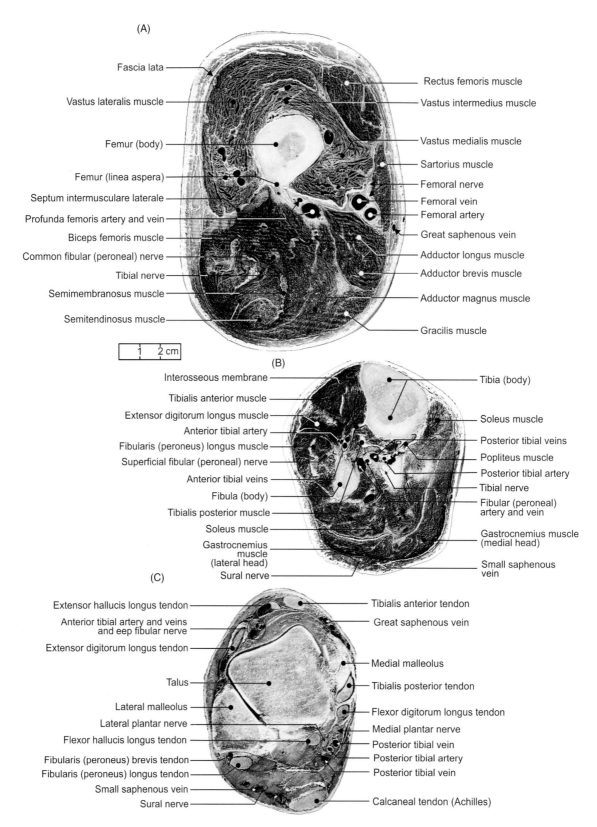

(A)

Fascia lata

Vastus lateralis muscle

Femur (body)

Femur (linea aspera)

Septum intermusculare laterale

Profunda femoris artery and vein

Biceps femoris muscle

Common fibular (peroneal) nerve

Tibial nerve

Semimembranosus muscle

Semitendinosus muscle

Rectus femoris muscle

Vastus intermedius muscle

Vastus medialis muscle

Sartorius muscle

Femoral nerve

Femoral vein

Femoral artery

Great saphenous vein

Adductor longus muscle

Adductor brevis muscle

Adductor magnus muscle

Gracilis muscle

1   2 cm

(B)

Interosseous membrane

Tibialis anterior muscle

Extensor digitorum longus muscle

Anterior tibial artery

Fibularis (peroneus) longus muscle

Superficial fibular (peroneal) nerve

Anterior tibial veins

Fibula (body)

Tibialis posterior muscle

Soleus muscle

Gastrocnemius muscle (lateral head)

Sural nerve

Tibia (body)

Soleus muscle

Posterior tibial veins

Popliteus muscle

Posterior tibial artery

Tibial nerve

Fibular (peroneal) artery and vein

Gastrocnemius muscle (medial head)

Small saphenous vein

(C)

Extensor hallucis longus tendon

Anterior tibial artery and veins and eep fibular nerve

Extensor digitorum longus tendon

Talus

Lateral malleolus

Lateral plantar nerve

Flexor hallucis longus tendon

Fibularis (peroneus) brevis tendon

Fibularis (peroneus) longus tendon

Small saphenous vein

Sural nerve

Tibialis anterior tendon

Great saphenous vein

Medial malleolus

Tibialis posterior tendon

Flexor digitorum longus tendon

Medial plantar nerve

Posterior tibial vein

Posterior tibial artery

Posterior tibial vein

Calcaneal tendon (Achilles)

FIGURE 2.24 **The right limb.** Cross-sectional views of the (A) thigh, (B) leg, and (C) proximal segment of foot.

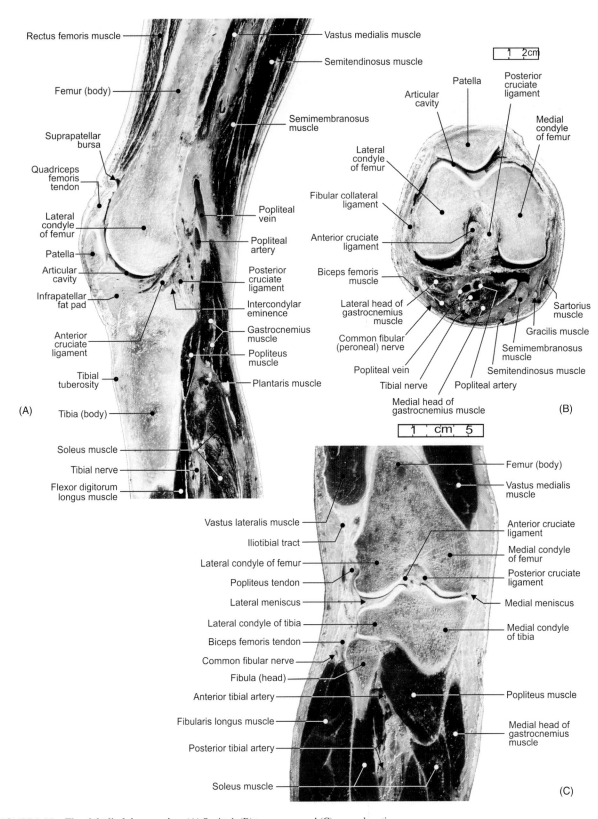

Rectus femoris muscle

Femur (body)

Suprapatellar bursa

Quadriceps femoris tendon

Lateral condyle of femur

Patella

Articular cavity

Infrapatellar fat pad

Anterior cruciate ligament

Tibial tuberosity

Tibia (body)

Soleus muscle

Tibial nerve

Flexor digitorum longus muscle

Vastus medialis muscle

Semitendinosus muscle

Semimembranosus muscle

Popliteal vein

Popliteal artery

Posterior cruciate ligament

Intercondylar eminence

Gastrocnemius muscle

Popliteus muscle

Plantaris muscle

(A)

1 2cm

Articular cavity

Patella

Lateral condyle of femur

Fibular collateral ligament

Anterior cruciate ligament

Biceps femoris muscle

Lateral head of gastrocnemius muscle

Common fibular (peroneal) nerve

Popliteal vein

Tibial nerve

Medial head of gastrocnemius muscle

Posterior cruciate ligament

Medial condyle of femur

Sartorius muscle

Gracilis muscle

Semimembranosus muscle

Semitendinosus muscle

Popliteal artery

(B)

1   cm 5

Vastus lateralis muscle

Iliotibial tract

Lateral condyle of femur

Popliteus tendon

Lateral meniscus

Lateral condyle of tibia

Biceps femoris tendon

Common fibular nerve

Fibula (head)

Anterior tibial artery

Fibularis longus muscle

Posterior tibial artery

Soleus muscle

Femur (body)

Vastus medialis muscle

Anterior cruciate ligament

Medial condyle of femur

Posterior cruciate ligament

Medial meniscus

Medial condyle of tibia

Popliteus muscle

Medial head of gastrocnemius muscle

(C)

FIGURE 2.25   The right limb knee region. (A) Sagittal, (B) transverse, and (C) coronal sections.

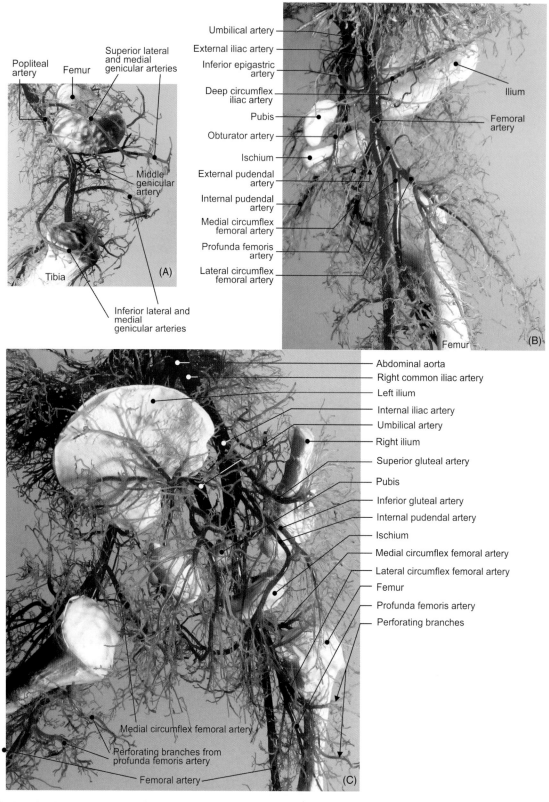

FIGURE 2.26 **The fetal left limb.** Arterial distribution (corrosion casts) around the (A) knee region, (B) anterior aspect of thigh, and (C) gluteal region.

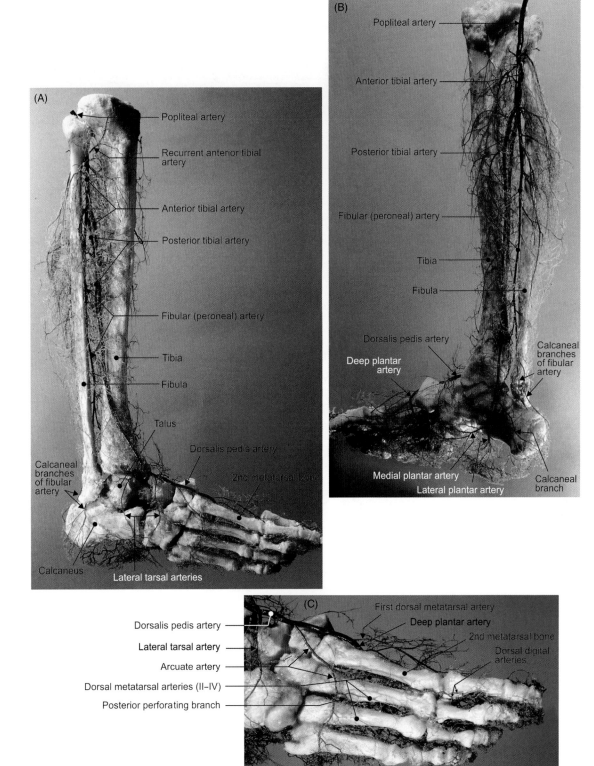

FIGURE 2.27 Arterial distributions (corrosion cast) along the (A) lateral and (B) medial aspects of the right leg and (C) dorsal aspect of the right foot.

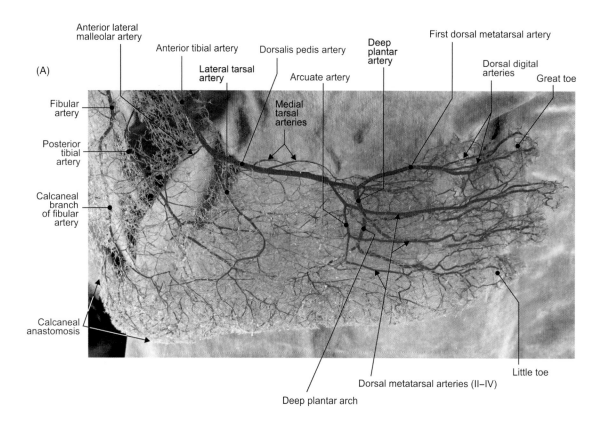

(A)

Anterior lateral malleolar artery

Anterior tibial artery

Lateral tarsal artery

Dorsalis pedis artery

Deep plantar artery

First dorsal metatarsal artery

Dorsal digital arteries

Great toe

Fibular artery

Arcuate artery

Medial tarsal arteries

Posterior tibial artery

Calcaneal branch of fibular artery

Calcaneal anastomosis

Little toe

Dorsal metatarsal arteries (II–IV)

Deep plantar arch

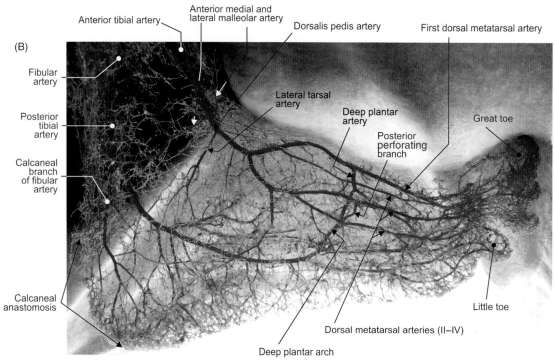

(B)

Anterior tibial artery

Anterior medial and lateral malleolar artery

Dorsalis pedis artery

First dorsal metatarsal artery

Fibular artery

Lateral tarsal artery

Deep plantar artery

Posterior perforating branch

Great toe

Posterior tibial artery

Calcaneal branch of fibular artery

Calcaneal anastomosis

Little toe

Dorsal metatarsal arteries (II–IV)

Deep plantar arch

FIGURE 2.28 (A–B) Different specimens depicting arterial distributions (corrosion casts) along the dorsal aspect of the right foot.

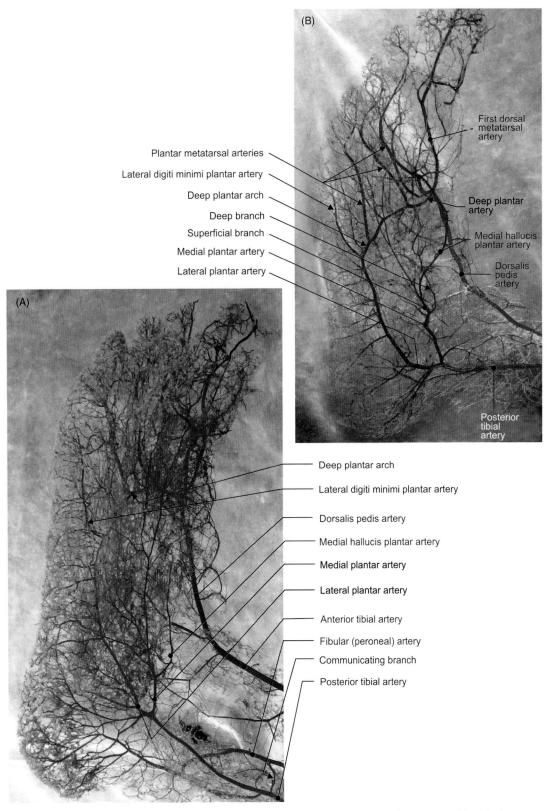

First dorsal metatarsal artery

Deep plantar artery

Medial hallucis plantar artery

Dorsalis pedis artery

Posterior tibial artery

Plantar metatarsal arteries

Lateral digiti minimi plantar artery

Deep plantar arch

Deep branch

Superficial branch

Medial plantar artery

Lateral plantar artery

Deep plantar arch

Lateral digiti minimi plantar artery

Dorsalis pedis artery

Medial hallucis plantar artery

Medial plantar artery

Lateral plantar artery

Anterior tibial artery

Fibular (peroneal) artery

Communicating branch

Posterior tibial artery

FIGURE 2.29 (A–B) Different specimens depicting arterial distributions (corrosion casts) along the plantar aspect of the right foot.

Chapter 3

Thorax and Vascularization

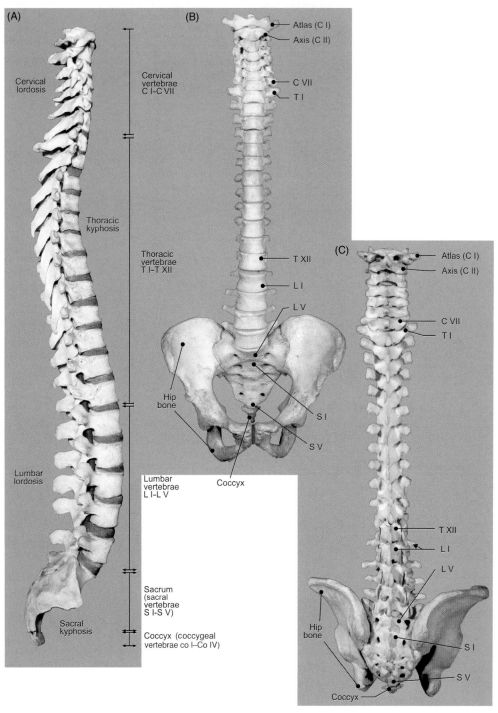

FIGURE 3.1 **The vertebral column and bony pelvis.** (A) Right, (B) anterior, and (C) posterior aspects.

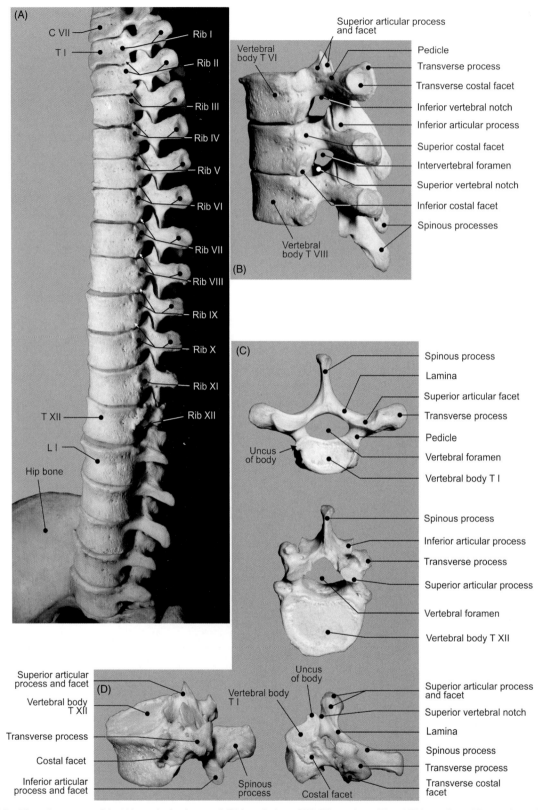

FIGURE 3.2 Thoracic segment of the (A) vertebral column and (B) lateral view of T6–T8 vertebrae. T1 and T12 vertebrae: (C) superior and (D) lateral views.

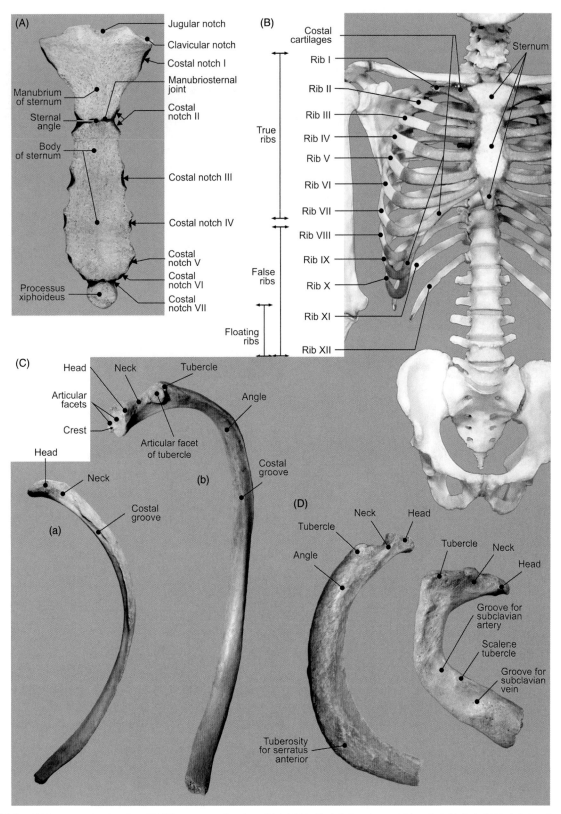

FIGURE 3.3   (A) Sternum: anterior view. (B) Rib cage: anterior view. (C) (a) 11th rib: inferior view and (b) typical rib (3rd–10th): inferior view. (D) 1st and 2nd right ribs: superior view.

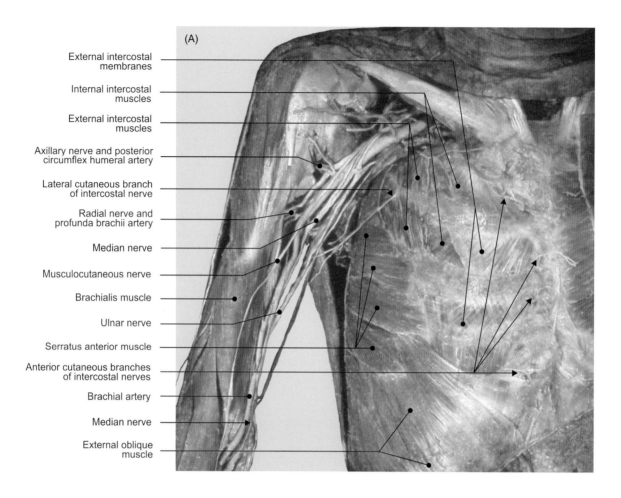

External intercostal membranes

Internal intercostal muscles

External intercostal muscles

Axillary nerve and posterior circumflex humeral artery

Lateral cutaneous branch of intercostal nerve

Radial nerve and profunda brachii artery

Median nerve

Musculocutaneous nerve

Brachialis muscle

Ulnar nerve

Serratus anterior muscle

Anterior cutaneous branches of intercostal nerves

Brachial artery

Median nerve

External oblique muscle

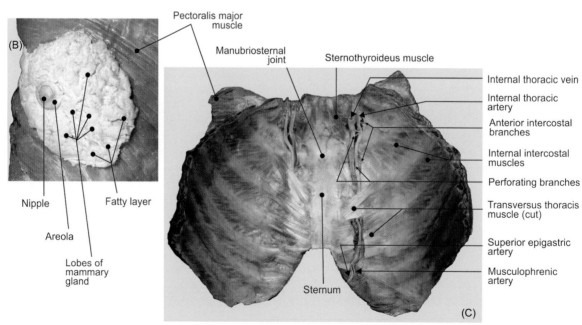

Pectoralis major muscle

Manubriosternal joint

Sternothyroideus muscle

Internal thoracic vein

Internal thoracic artery

Anterior intercostal branches

Internal intercostal muscles

Perforating branches

Transversus thoracis muscle (cut)

Superior epigastric artery

Musculophrenic artery

Nipple

Fatty layer

Areola

Lobes of mammary gland

Sternum

FIGURE 3.4   (A) Pectoral and axillary regions. (B) Dissected mammary gland. (C) Thoracic cage: posterior aspect of the anterior segment.

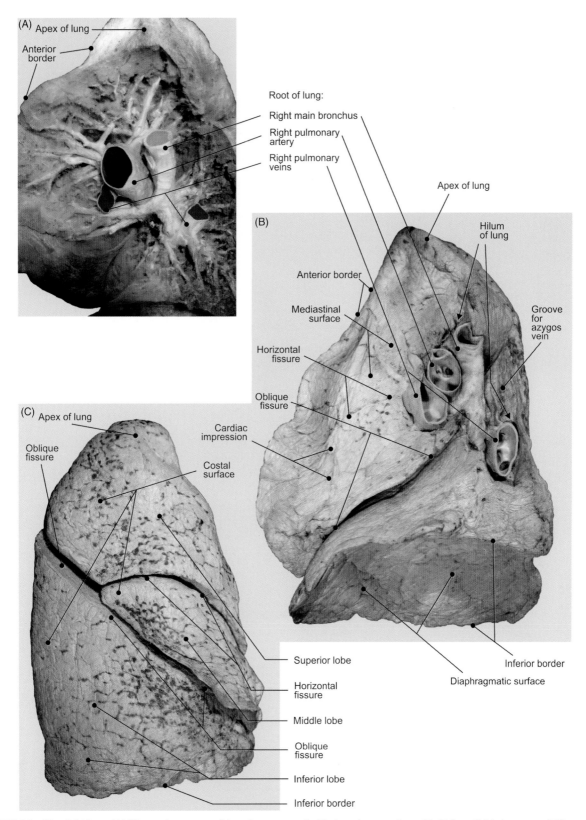

FIGURE 3.5 **The right lung.** (A) Dissected structures of the pulmonary root inside the pulmonary tissue. (B–C) Superficial elements and (C) structures of the pulmonary root.

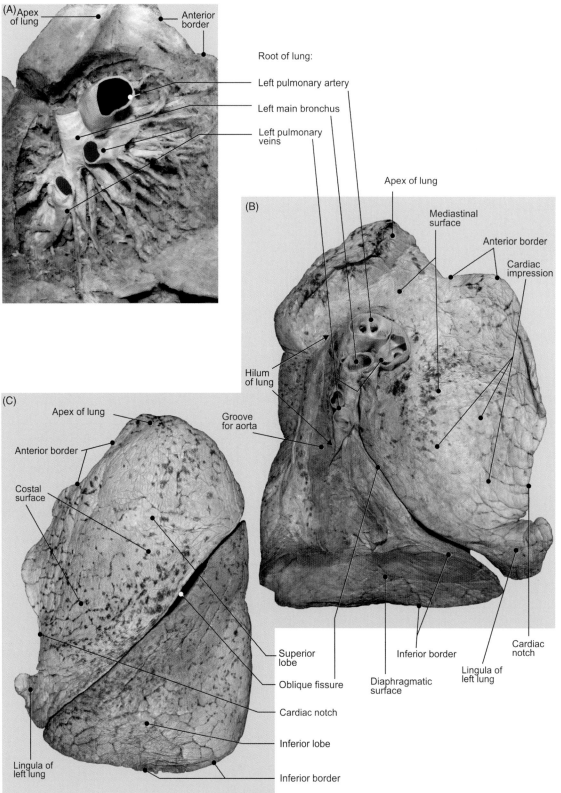

FIGURE 3.6 The left lung. (A) Dissected structures of the pulmonary root inside the pulmonary tissue. (B–C) Superficial elements and (C) structures of the pulmonary root.

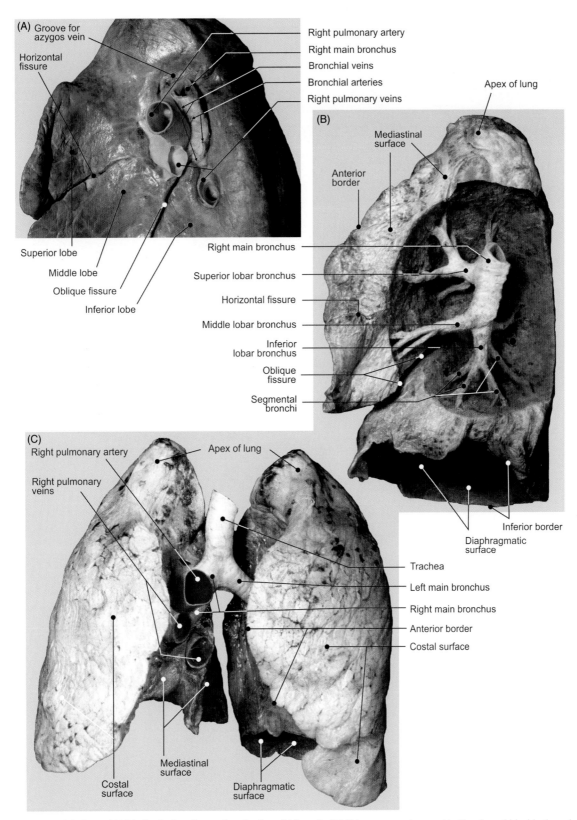

FIGURE 3.7   The right lung. (A) Mediastinal surface and tracheobronchial trunk. (B) Primary, secondary, and tertiary bronchi inside the pulmonary tissue. (C) The anterior aspect of lungs, trachea, and bronchi.

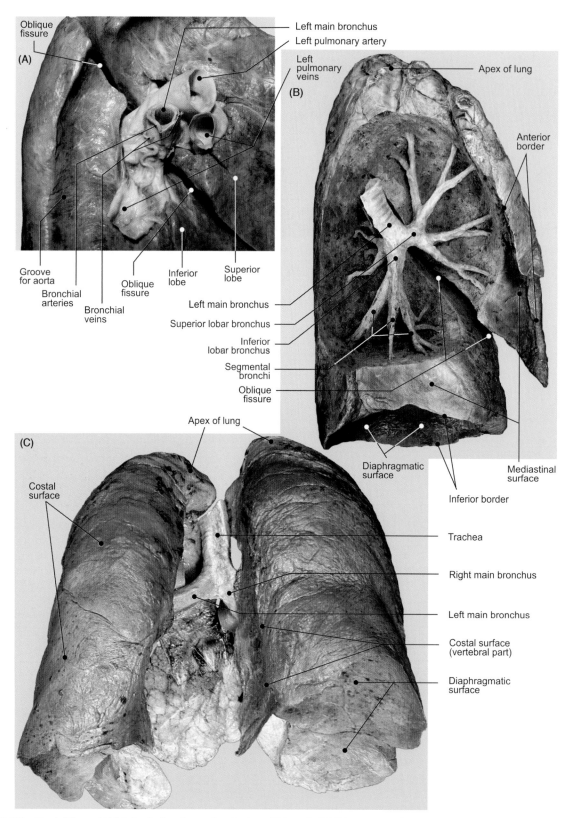

FIGURE 3.8   The left lung. (A) Mediastinal surface and tracheobronchial trunk. (B) Primary, secondary, and tertiary bronchi inside the pulmonary tissue. (C) Posterior aspect of lungs, trachea, and bronchi.

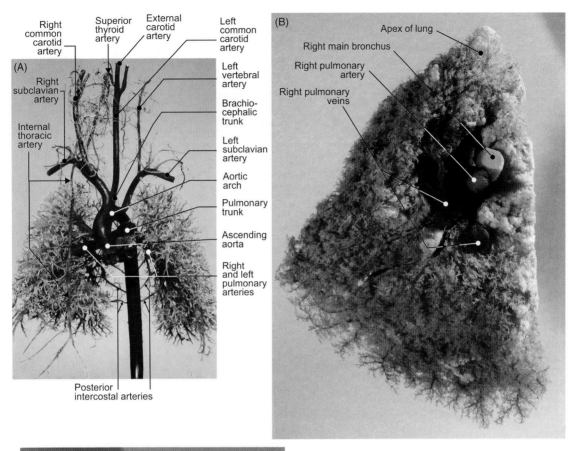

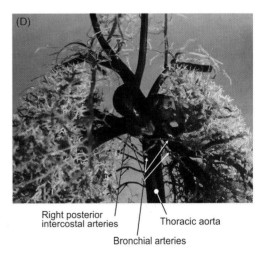

FIGURE 3.9 (A) Branching of pulmonary arteries and aorta (corrosion cast, fetus). (B–C) Right lung, corrosion cast: branching of the pulmonary artery *(blue)*, pulmonary veins *(red)*, and bronchial trunk *(yellow)*; (B) mediastinal surface and (C) costal surface. (D) Bronchial arteries (corrosion cast, fetus).

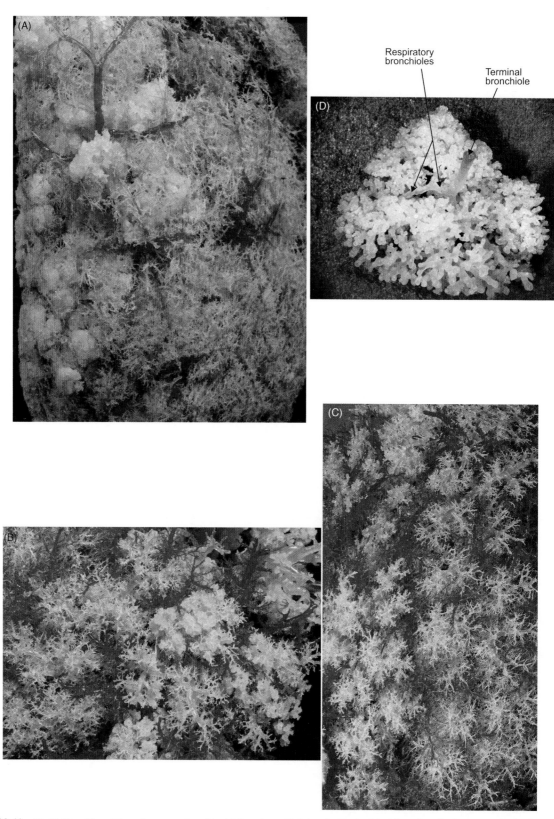

FIGURE 3.10 (A–C) Branching of the pulmonary artery *(blue)* follows bronchial stem *(yellow)*, while tributaries of pulmonary veins *(red)* are peripheral (corrosion cast). (D) Terminal bronchiole continues as respiratory bronchiole, alveolar sac, and, eventually, pulmonary alveoli (corrosion cast).

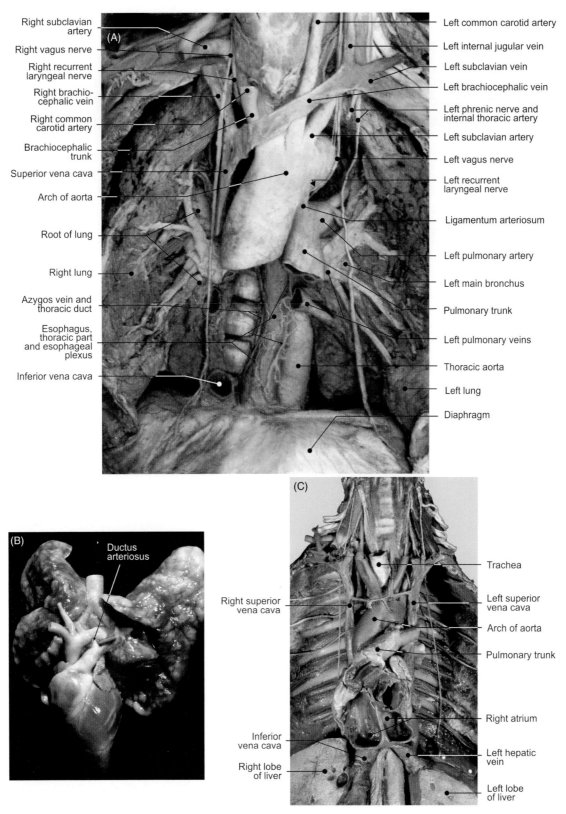

Right subclavian artery
Right vagus nerve
Right recurrent laryngeal nerve
Right brachio-cephalic vein
Right common carotid artery
Brachiocephalic trunk
Superior vena cava
Arch of aorta
Root of lung
Right lung
Azygos vein and thoracic duct
Esophagus, thoracic part and esophageal plexus
Inferior vena cava

Left common carotid artery
Left internal jugular vein
Left subclavian vein
Left brachiocephalic vein
Left phrenic nerve and internal thoracic artery
Left subclavian artery
Left vagus nerve
Left recurrent laryngeal nerve
Ligamentum arteriosum
Left pulmonary artery
Left main bronchus
Pulmonary trunk
Left pulmonary veins
Thoracic aorta
Left lung
Diaphragm

(B)
Ductus arteriosus

(C)
Trachea
Left superior vena cava
Arch of aorta
Pulmonary trunk
Right atrium
Left hepatic vein
Left lobe of liver

Right superior vena cava
Inferior vena cava
Right lobe of liver

FIGURE 3.11   (A) The anterior aspect of lungs and mediastinum in situ after the heart has been removed (layer of blood vessels). (B) The anterior aspect of fetal lungs and heart. (C) Double superior caval vein.

(A)

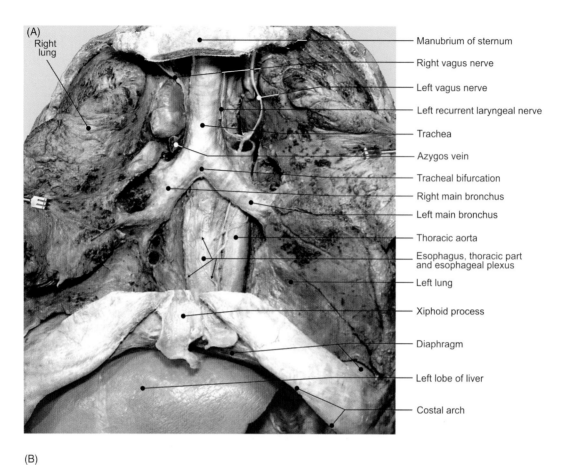

Right lung

Manubrium of sternum

Right vagus nerve

Left vagus nerve

Left recurrent laryngeal nerve

Trachea

Azygos vein

Tracheal bifurcation

Right main bronchus

Left main bronchus

Thoracic aorta

Esophagus, thoracic part and esophageal plexus

Left lung

Xiphoid process

Diaphragm

Left lobe of liver

Costal arch

(B)

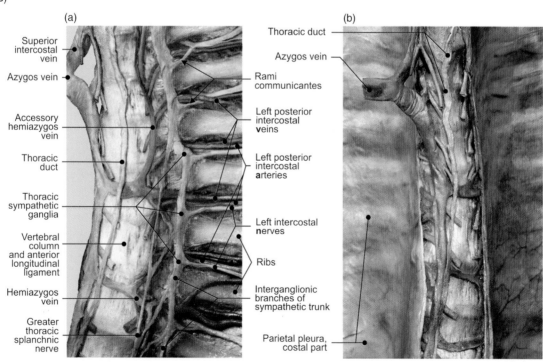

(a)

Superior intercostal vein

Azygos vein

Accessory hemiazygos vein

Thoracic duct

Thoracic sympathetic ganglia

Vertebral column and anterior longitudinal ligament

Hemiazygos vein

Greater thoracic splanchnic nerve

Rami communicantes

Left posterior intercostal **v**eins

Left posterior intercostal **a**rteries

Left intercostal **n**erves

Ribs

Interganglionic branches of sympathetic trunk

(b)

Thoracic duct

Azygos vein

Parietal pleura, costal part

FIGURE 3.12 (A) The anterior aspect of lungs and mediastinum in situ. The major blood vessels and heart have been removed. (B) Posterior mediastinum (a) with blood vessels and nerves and (b) after removal of costal pleura.

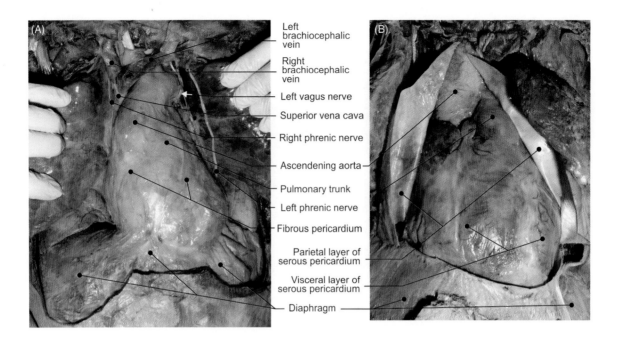

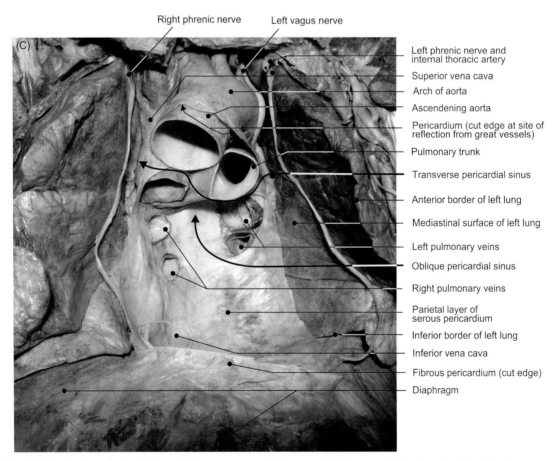

FIGURE 3.13 (A) Heart and pericardium in situ. (B) The anterior aspect of the heart. (C) The pericardial cavity after the heart has been removed.

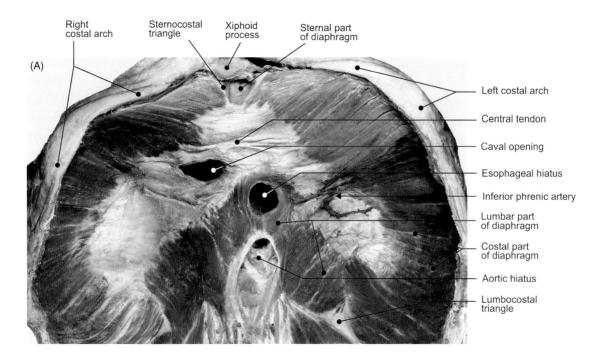

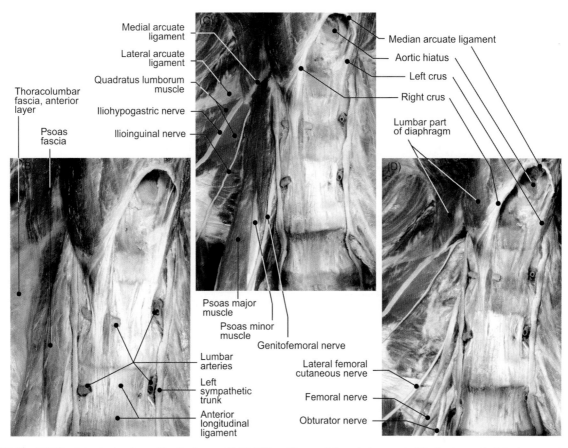

FIGURE 3.14   (A) The abdominal surface of the diaphragm. (B–D) Different layers of the anterior aspect of the lumbar part of diaphragm and parts of the posterior abdominal wall.

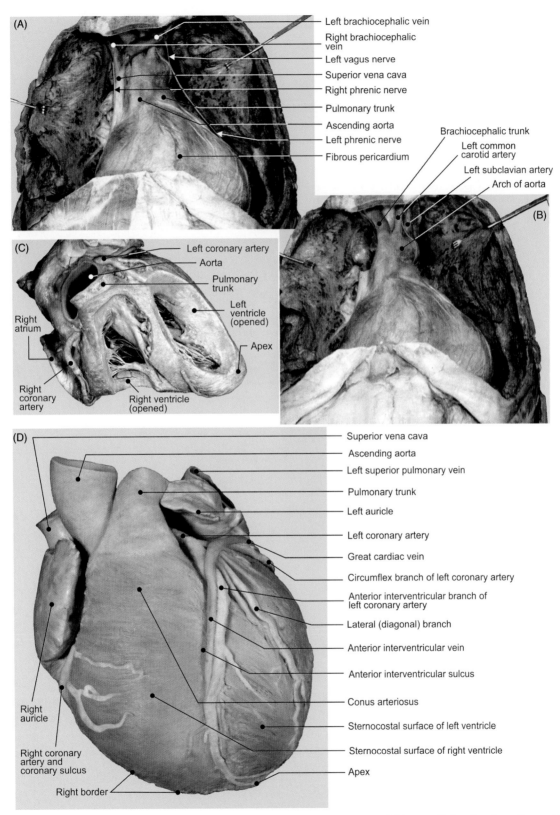

FIGURE 3.15 Superior mediastinum and heart in pericardium in situ: (A) venous and (B) arterial layers. (C) Anterior view of heart with opened ventricles. (D) Anterior view of heart.

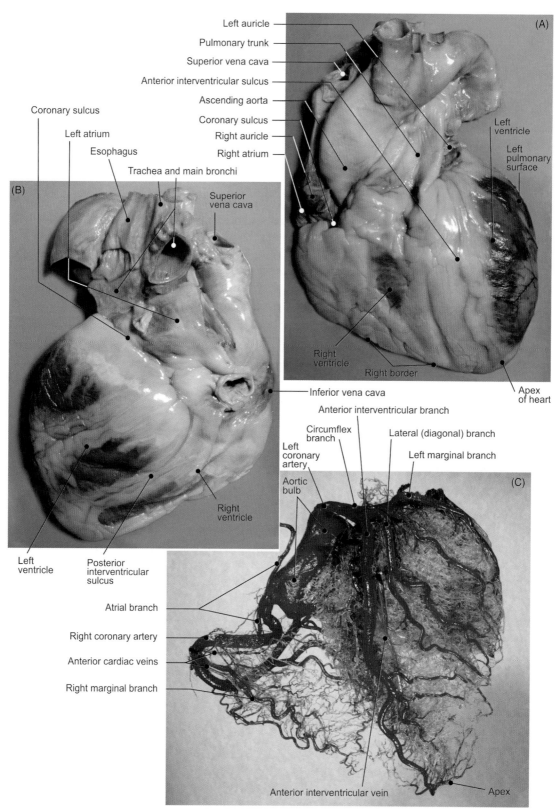

FIGURE 3.16 The native heart: (A) sternocostal surface and (B) diaphragmatic surface. (C) Anterior view: corrosion cast of cardiac arteries *(red)* and veins *(blue)*.

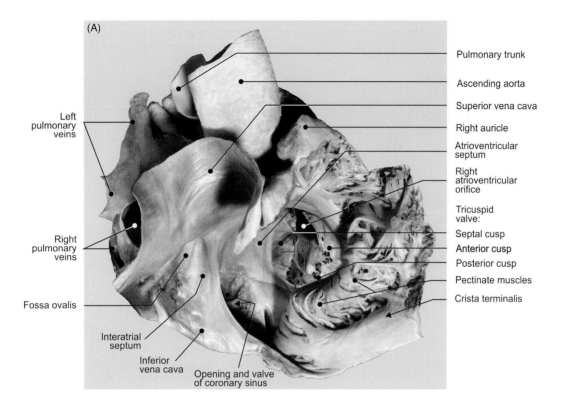

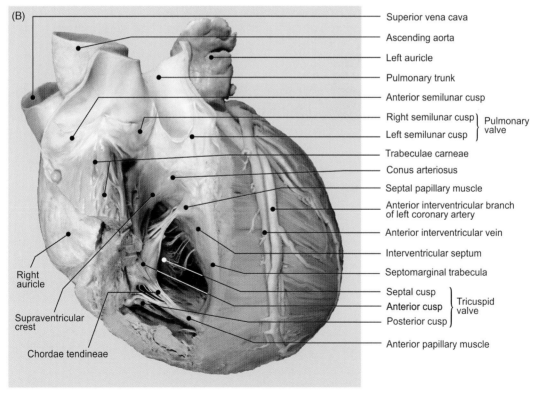

FIGURE 3.17   (A) The right atrium: view from above. (B) The right ventricle: view from the front.

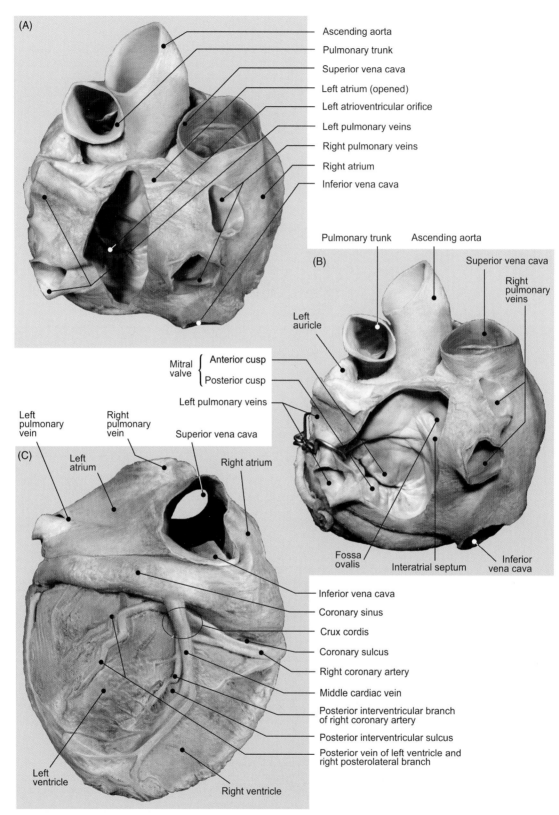

(A)

Ascending aorta
Pulmonary trunk
Superior vena cava
Left atrium (opened)
Left atrioventricular orifice
Left pulmonary veins
Right pulmonary veins
Right atrium
Inferior vena cava

(B)

Pulmonary trunk   Ascending aorta
Superior vena cava
Right pulmonary veins
Left auricle

Mitral valve { Anterior cusp
Posterior cusp }
Left pulmonary veins
Superior vena cava

Fossa ovalis   Interatrial septum   Inferior vena cava

(C)

Left pulmonary vein   Right pulmonary vein   Superior vena cava
Left atrium   Right atrium

Inferior vena cava
Coronary sinus
Crux cordis
Coronary sulcus
Right coronary artery
Middle cardiac vein
Posterior interventricular branch of right coronary artery
Posterior interventricular sulcus
Posterior vein of left ventricle and right posterolateral branch

Left ventricle   Right ventricle

FIGURE 3.18   Cardiac base: (A) view from behind and (B) with open posterior wall. (C) Diaphragmatic surface of the heart.

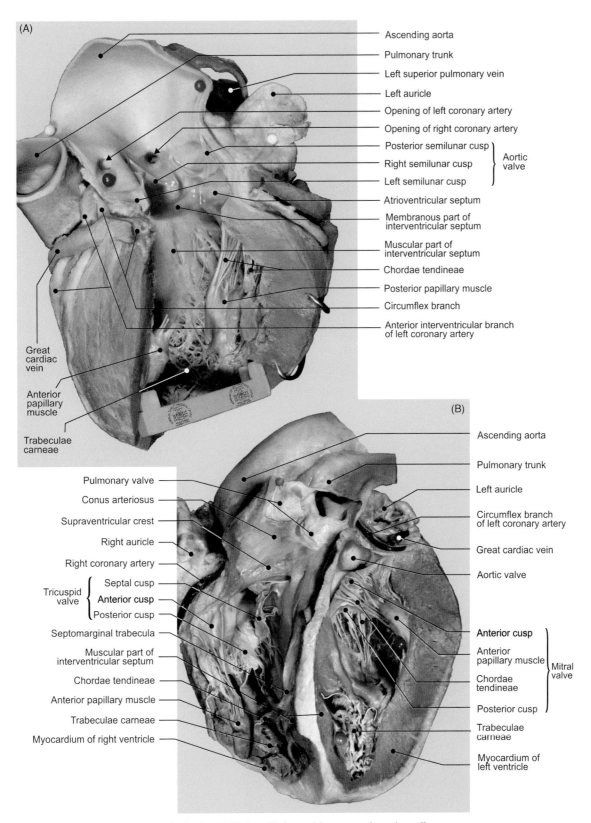

(A)

Ascending aorta

Pulmonary trunk

Left superior pulmonary vein

Left auricle

Opening of left coronary artery

Opening of right coronary artery

Posterior semilunar cusp ⎱
Right semilunar cusp   ⎰ Aortic valve
Left semilunar cusp

Atrioventricular septum

Membranous part of interventricular septum

Muscular part of interventricular septum

Chordae tendineae

Posterior papillary muscle

Circumflex branch

Anterior interventricular branch of left coronary artery

Great cardiac vein

Anterior papillary muscle

Trabeculae carneae

(B)

Pulmonary valve

Conus arteriosus

Supraventricular crest

Right auricle

Right coronary artery

Tricuspid valve { Septal cusp
Anterior cusp
Posterior cusp }

Septomarginal trabecula

Muscular part of interventricular septum

Chordae tendineae

Anterior papillary muscle

Trabeculae carneae

Myocardium of right ventricle

Ascending aorta

Pulmonary trunk

Left auricle

Circumflex branch of left coronary artery

Great cardiac vein

Aortic valve

Anterior cusp ⎱
Anterior papillary muscle   ⎰ Mitral valve
Chordae tendineae
Posterior cusp

Trabeculae carneae

Myocardium of left ventricle

FIGURE 3.19   (A) Left ventricle: open longitudinally. (B) Right and left ventricles: removed anterior wall.

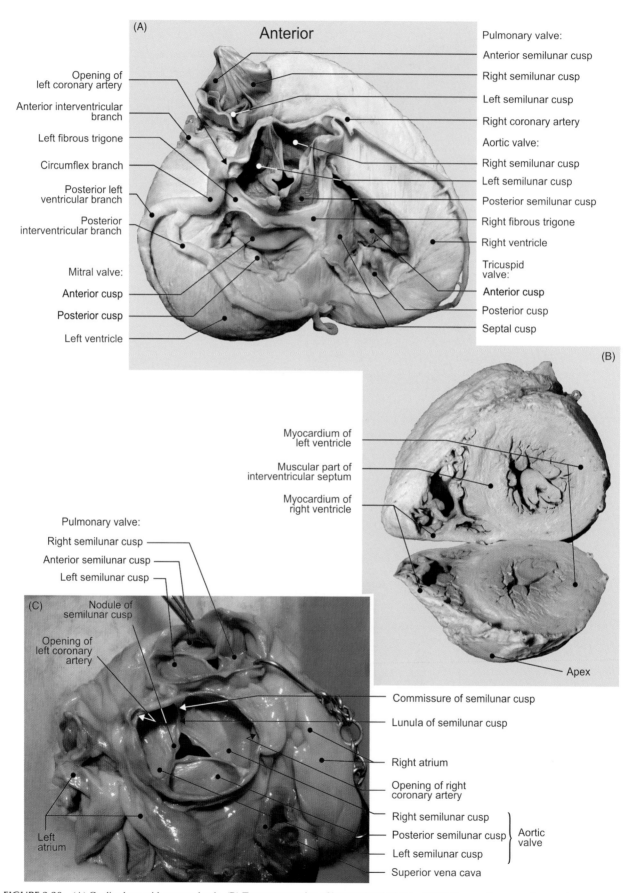

(A)

Anterior

Opening of
left coronary artery

Anterior interventricular
branch

Left fibrous trigone

Circumflex branch

Posterior left
ventricular branch

Posterior
interventricular branch

Mitral valve:

Anterior cusp

Posterior cusp

Left ventricle

Pulmonary valve:

Anterior semilunar cusp

Right semilunar cusp

Left semilunar cusp

Right coronary artery

Aortic valve:

Right semilunar cusp

Left semilunar cusp

Posterior semilunar cusp

Right fibrous trigone

Right ventricle

Tricuspid
valve:

Anterior cusp

Posterior cusp

Septal cusp

(B)

Myocardium of
left ventricle

Muscular part of
interventricular septum

Myocardium of
right ventricle

Apex

Pulmonary valve:

Right semilunar cusp

Anterior semilunar cusp

Left semilunar cusp

(C)

Nodule of
semilunar cusp

Opening of
left coronary
artery

Left
atrium

Commissure of semilunar cusp

Lunula of semilunar cusp

Right atrium

Opening of right
coronary artery

Right semilunar cusp

Posterior semilunar cusp   } Aortic
valve

Left semilunar cusp

Superior vena cava

FIGURE 3.20   (A) Cardiac base with removed atria. (B) Transverse section of heart at level of apex. (C) Native specimen: semilunar valves.

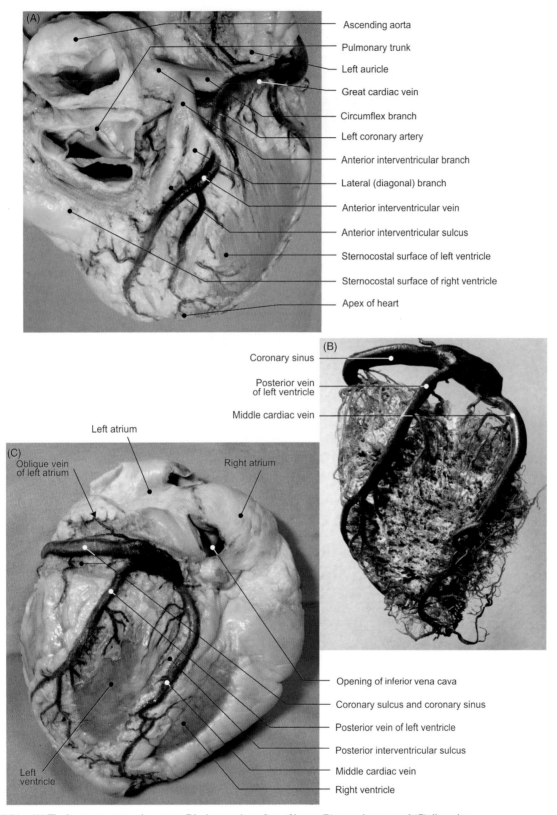

FIGURE 3.21   (A) The heart: anterosuperior aspect. Diaphragmatic surface of heart: (B) corrosion cast and (C) dissection.

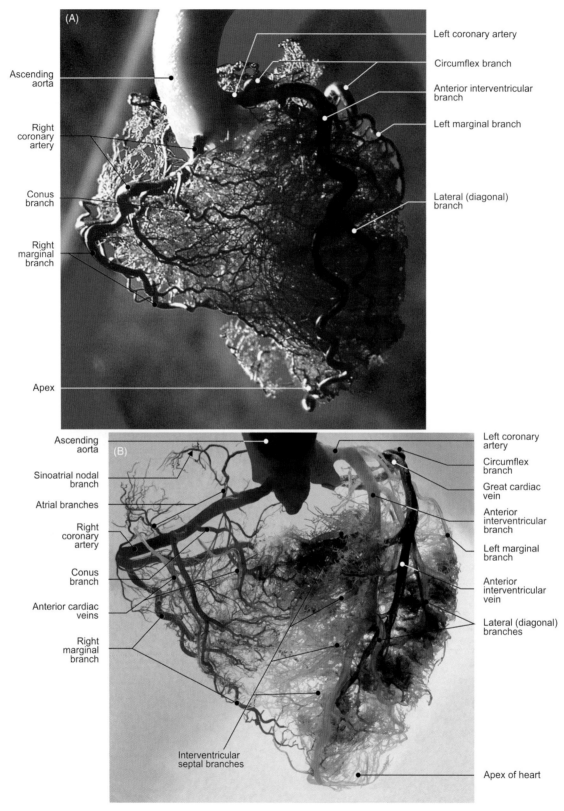

FIGURE 3.22   Sternocostal surface of the heart. (A–B) Different specimens of the arterial and venous distributions (corrosion casts).

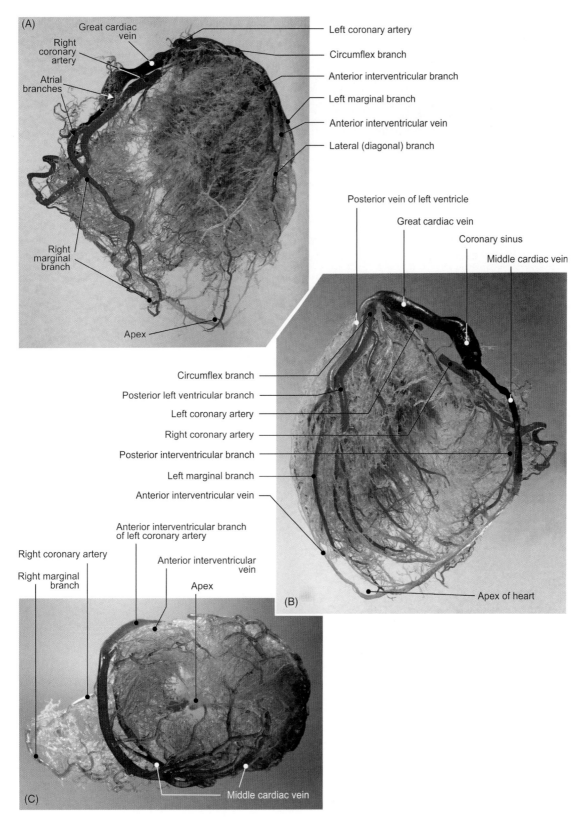

FIGURE 3.23 Cardiac arterial and venous distributions (corrosion cast). (A) Sternocostal surface. (B) Diaphragmatic surface (adjusted tip of circulation). (C) Inferior view of the cardiac apex.

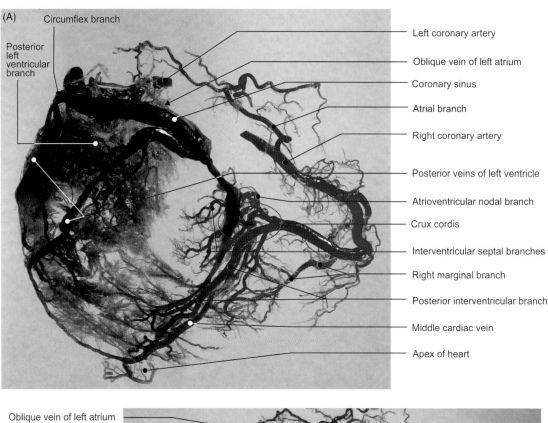

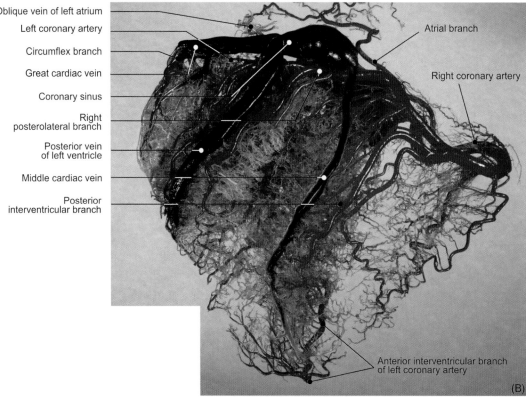

FIGURE 3.24 **Cardiac arterial and venous distributions (diaphragmatic surfaces of corrosion casts).** (A) Adjusted tip of circulation and (B) dominant right coronary artery (specimen from Fig. 3.16).

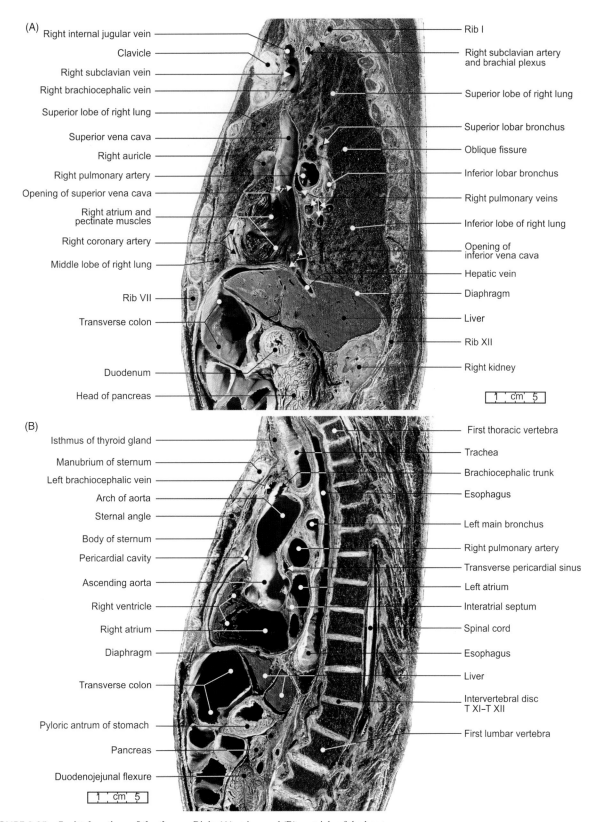

(A)
Right internal jugular vein
Clavicle
Right subclavian vein
Right brachiocephalic vein
Superior lobe of right lung
Superior vena cava
Right auricle
Right pulmonary artery
Opening of superior vena cava
Right atrium and pectinate muscles
Right coronary artery
Middle lobe of right lung
Rib VII
Transverse colon
Duodenum
Head of pancreas

Rib I
Right subclavian artery and brachial plexus
Superior lobe of right lung
Superior lobar bronchus
Oblique fissure
Inferior lobar bronchus
Right pulmonary veins
Inferior lobe of right lung
Opening of inferior vena cava
Hepatic vein
Diaphragm
Liver
Rib XII
Right kidney

1 cm 5

(B)
Isthmus of thyroid gland
Manubrium of sternum
Left brachiocephalic vein
Arch of aorta
Sternal angle
Body of sternum
Pericardial cavity
Ascending aorta
Right ventricle
Right atrium
Diaphragm
Transverse colon
Pyloric antrum of stomach
Pancreas
Duodenojejunal flexure

First thoracic vertebra
Trachea
Brachiocephalic trunk
Esophagus
Left main bronchus
Right pulmonary artery
Transverse pericardial sinus
Left atrium
Interatrial septum
Spinal cord
Esophagus
Liver
Intervertebral disc T XI–T XII
First lumbar vertebra

1 cm 5

FIGURE 3.25   Sagittal sections of the thorax. Right (A) atrium and (B) ventricle of the heart.

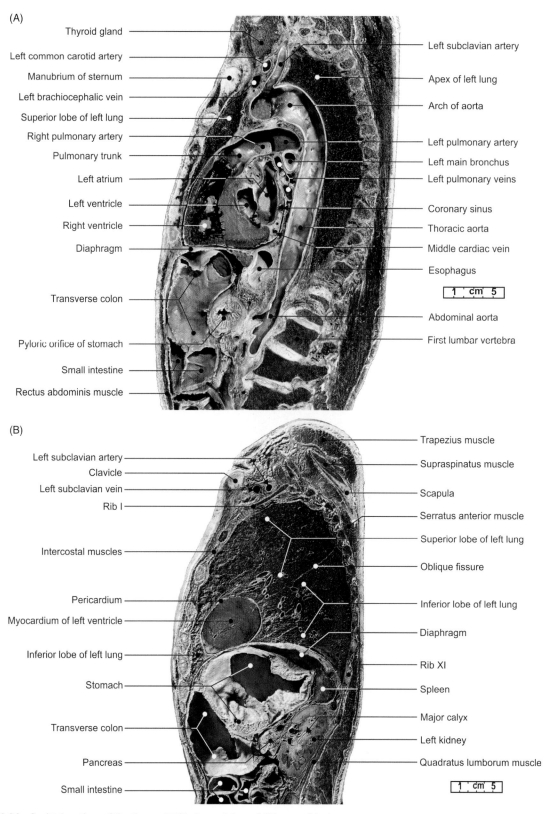

(A)

Thyroid gland
Left common carotid artery
Manubrium of sternum
Left brachiocephalic vein
Superior lobe of left lung
Right pulmonary artery
Pulmonary trunk
Left atrium
Left ventricle
Right ventricle
Diaphragm

Transverse colon

Pyloric orifice of stomach

Small intestine

Rectus abdominis muscle

Left subclavian artery
Apex of left lung
Arch of aorta
Left pulmonary artery
Left main bronchus
Left pulmonary veins
Coronary sinus
Thoracic aorta
Middle cardiac vein
Esophagus

1 cm 5

Abdominal aorta
First lumbar vertebra

(B)

Left subclavian artery
Clavicle
Left subclavian vein
Rib I

Intercostal muscles

Pericardium
Myocardium of left ventricle

Inferior lobe of left lung

Stomach

Transverse colon

Pancreas

Small intestine

Trapezius muscle
Supraspinatus muscle
Scapula
Serratus anterior muscle
Superior lobe of left lung
Oblique fissure
Inferior lobe of left lung
Diaphragm
Rib XI
Spleen
Major calyx
Left kidney
Quadratus lumborum muscle

1 cm 5

FIGURE 3.26 Sagittal sections of the thorax. (A) Both ventricles and (B) apex of the heart.

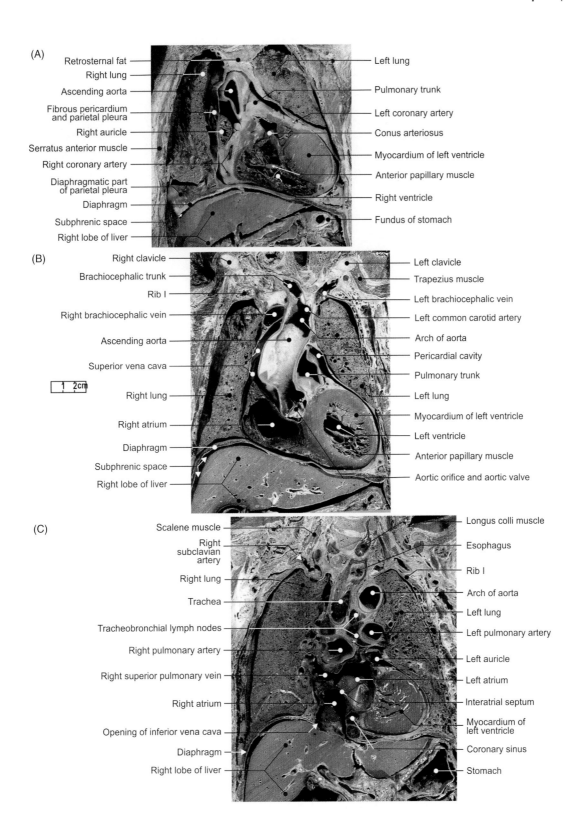

FIGURE 3.27   Coronal sections of the thorax. (A) Right ventricle, (B) left ventricle, and (C) left atrium.

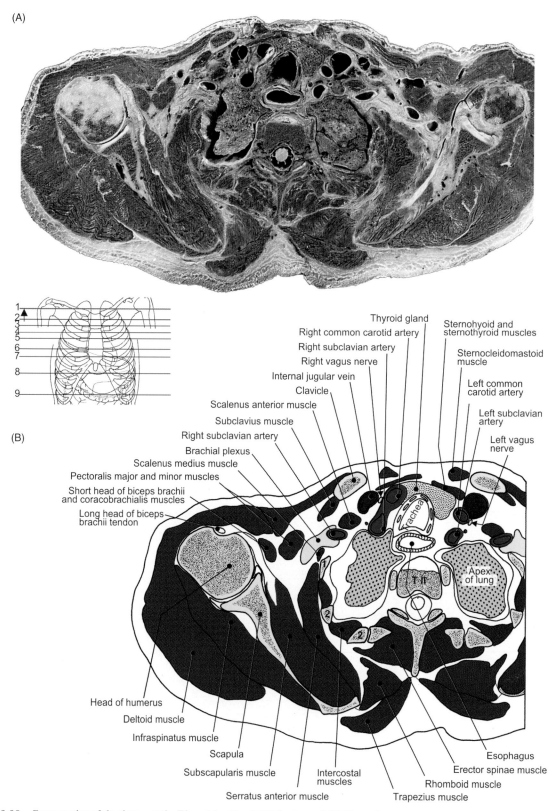

(A)

(B)

Thyroid gland
Right common carotid artery
Right subclavian artery
Right vagus nerve
Internal jugular vein
Clavicle
Scalenus anterior muscle
Subclavius muscle
Right subclavian artery
Brachial plexus
Scalenus medius muscle
Pectoralis major and minor muscles
Short head of biceps brachii and coracobrachialis muscles
Long head of biceps brachii tendon

Sternohyoid and sternothyroid muscles
Sternocleidomastoid muscle
Left common carotid artery
Left subclavian artery
Left vagus nerve

Trachea
Apex of lung
T II

Head of humerus
Deltoid muscle
Infraspinatus muscle
Scapula
Subscapularis muscle
Serratus anterior muscle
Intercostal muscles
Esophagus
Erector spinae muscle
Rhomboid muscle
Trapezius muscle

FIGURE 3.28   Cross-section of the thorax at the T2 vertebral level. (A) Section and (B) illustration.

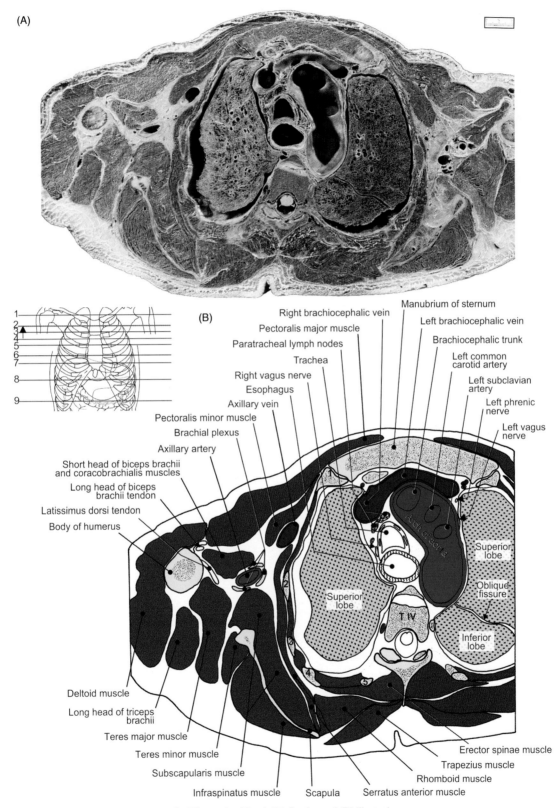

(A)

(B)

Manubrium of sternum
Right brachiocephalic vein
Left brachiocephalic vein
Pectoralis major muscle
Brachiocephalic trunk
Paratracheal lymph nodes
Left common carotid artery
Trachea
Left subclavian artery
Right vagus nerve
Esophagus
Left phrenic nerve
Axillary vein
Left vagus nerve
Pectoralis minor muscle
Brachial plexus
Axillary artery
Short head of biceps brachii and coracobrachialis muscles
Long head of biceps brachii tendon
Latissimus dorsi tendon
Body of humerus

Superior lobe
Oblique fissure
Superior lobe
T IV
Inferior lobe
Arch of aorta

Deltoid muscle
Long head of triceps brachii
Teres major muscle
Teres minor muscle
Subscapularis muscle
Infraspinatus muscle
Scapula
Serratus anterior muscle
Erector spinae muscle
Trapezius muscle
Rhomboid muscle

FIGURE 3.29   Cross-section of the thorax at the T4 vertebral level. (A) Section and (B) illustration.

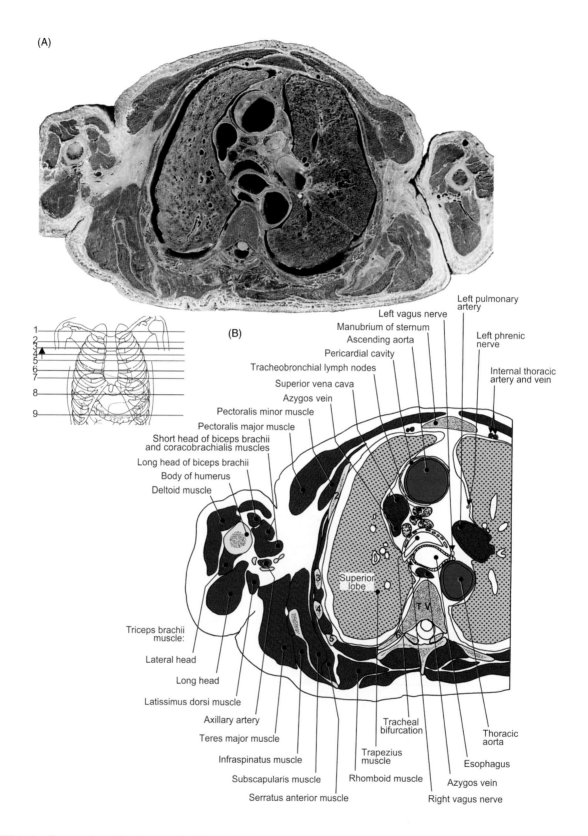

(A)

(B)

Left pulmonary artery

Left vagus nerve

Manubrium of sternum

Ascending aorta

Left phrenic nerve

Pericardial cavity

Tracheobronchial lymph nodes

Internal thoracic artery and vein

Superior vena cava

Azygos vein

Pectoralis minor muscle

Pectoralis major muscle

Short head of biceps brachii and coracobrachialis muscles

Long head of biceps brachii

Body of humerus

Deltoid muscle

Superior lobe

T V

Triceps brachii muscle:

Lateral head

Long head

Latissimus dorsi muscle

Axillary artery

Teres major muscle

Infraspinatus muscle

Subscapularis muscle

Serratus anterior muscle

Tracheal bifurcation

Trapezius muscle

Rhomboid muscle

Thoracic aorta

Esophagus

Azygos vein

Right vagus nerve

FIGURE 3.30   Cross-section of the thorax at the T5 vertebral level. (A) Section and (B) illustration.

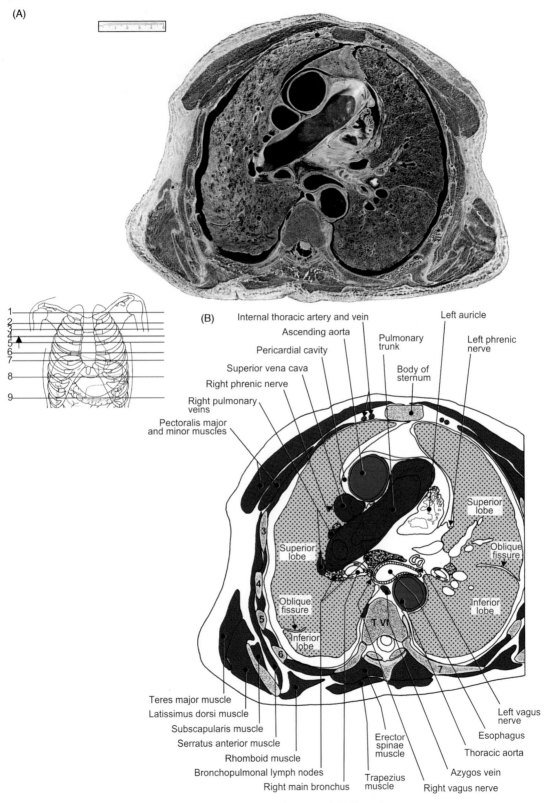

(A)

(B)

Internal thoracic artery and vein
Ascending aorta
Pericardial cavity
Superior vena cava
Right phrenic nerve
Right pulmonary veins
Pectoralis major and minor muscles

Left auricle
Pulmonary trunk
Body of sternum
Left phrenic nerve

Superior lobe
Superior lobe
Oblique fissure
Inferior lobe
Oblique fissure
Inferior lobe
T VI

Teres major muscle
Latissimus dorsi muscle
Subscapularis muscle
Serratus anterior muscle
Rhomboid muscle
Bronchopulmonal lymph nodes
Right main bronchus
Erector spinae muscle
Trapezius muscle
Right vagus nerve
Left vagus nerve
Esophagus
Thoracic aorta
Azygos vein

FIGURE 3.31 **Cross-section of the thorax at the T6 vertebral level.** (A) Section and (B) illustration.

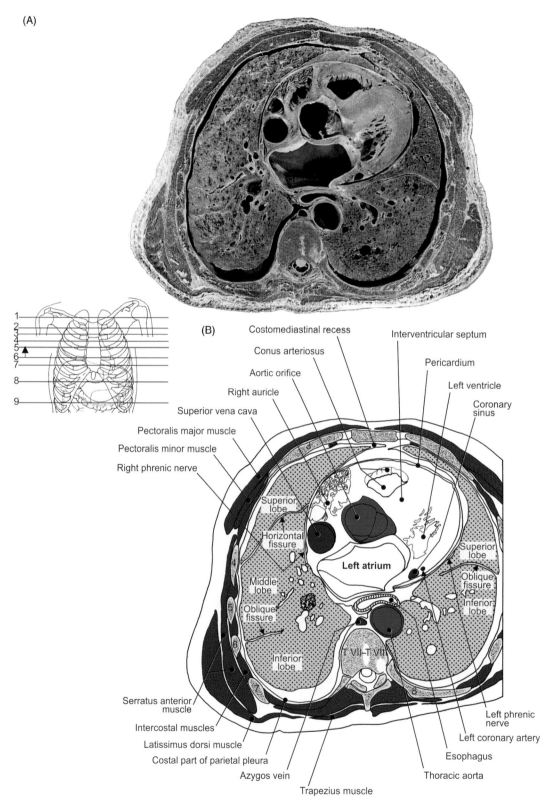

(A)

(B)

Costomediastinal recess

Conus arteriosus

Aortic orifice

Right auricle

Superior vena cava

Pectoralis major muscle

Pectoralis minor muscle

Right phrenic nerve

Interventricular septum

Pericardium

Left ventricle

Coronary sinus

Superior lobe

Horizontal fissure

Middle lobe

Oblique fissure

Inferior lobe

Left atrium

Superior lobe

Oblique fissure

Inferior lobe

T VII–T VIII

Serratus anterior muscle

Intercostal muscles

Latissimus dorsi muscle

Costal part of parietal pleura

Azygos vein

Trapezius muscle

Left phrenic nerve

Left coronary artery

Esophagus

Thoracic aorta

FIGURE 3.32 Cross-section of the thorax at the T7–T8 vertebral level. (A) Section and (B) illustration.

(A)

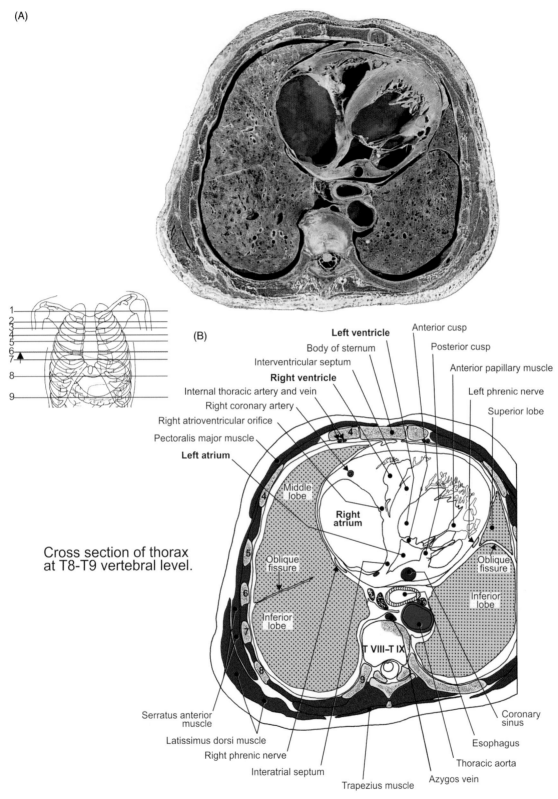

(B)

Cross section of thorax
at T8-T9 vertebral level.

Left ventricle
Body of sternum
Interventricular septum
Right ventricle
Internal thoracic artery and vein
Right coronary artery
Right atrioventricular orifice
Pectoralis major muscle
Left atrium

Anterior cusp
Posterior cusp
Anterior papillary muscle
Left phrenic nerve
Superior lobe

Middle lobe
Right atrium
Oblique fissure
Inferior lobe

Oblique fissure
Inferior lobe

T VIII–T IX

Serratus anterior muscle
Latissimus dorsi muscle
Right phrenic nerve
Interatrial septum
Trapezius muscle
Azygos vein
Thoracic aorta
Esophagus
Coronary sinus

FIGURE 3.33   Cross-section of the thorax at the T8–T9 vertebral level. (A) Section and (B) illustration.

(A)

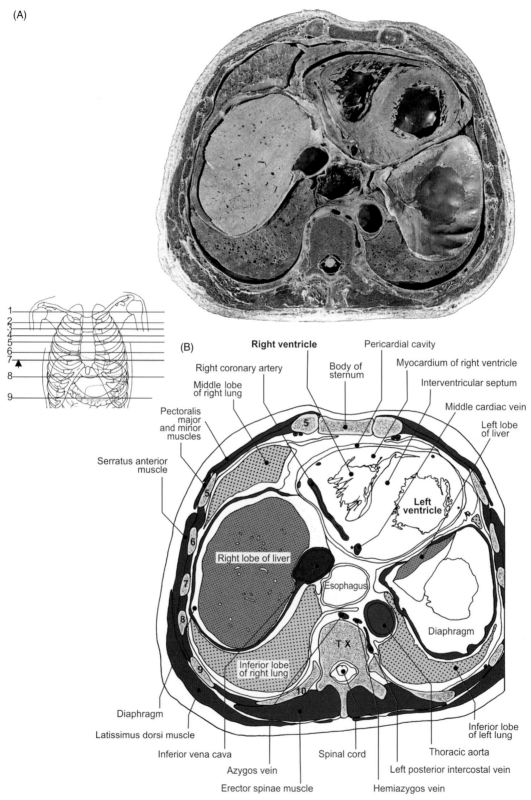

(B)

Right ventricle

Right coronary artery

Middle lobe
of right lung

Pectoralis
major
and minor
muscles

Serratus anterior
muscle

Right lobe of liver

Esophagus

Inferior lobe
of right lung

Diaphragm

Latissimus dorsi muscle

Inferior vena cava

Azygos vein

Erector spinae muscle

Body of
sternum

Pericardial cavity

Myocardium of right ventricle

Interventricular septum

Middle cardiac vein

Left lobe
of liver

Left
ventricle

Diaphragm

Inferior lobe
of left lung

Thoracic aorta

Left posterior intercostal vein

Hemiazygos vein

Spinal cord

FIGURE 3.34   Cross-section of the thorax at the T10 vertebral level. (A) Section and (B) illustration.

Chapter 4

Abdomen and Vascularization

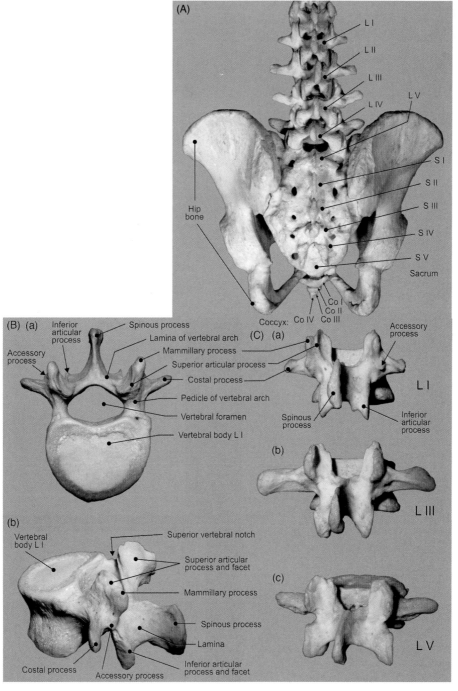

FIGURE 4.1 (A) Lumbar and sacral segments of the vertebral column: posterior view; total fusion of L5 vertebra with sacrum. (B) Vertebra L1: (a) superior and (b) lateral views. (C) Vertebrae posterior view: (a) L1, (b) L3, and (c) L5.

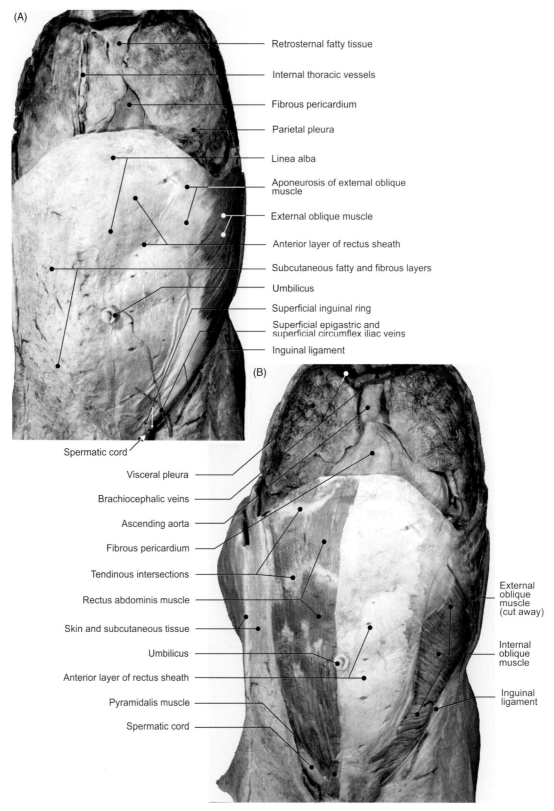

Retrosternal fatty tissue

Internal thoracic vessels

Fibrous pericardium

Parietal pleura

Linea alba

Aponeurosis of external oblique muscle

External oblique muscle

Anterior layer of rectus sheath

Subcutaneous fatty and fibrous layers

Umbilicus

Superficial inguinal ring

Superficial epigastric and superficial circumflex iliac veins

Inguinal ligament

Spermatic cord

Visceral pleura

Brachiocephalic veins

Ascending aorta

Fibrous pericardium

Tendinous intersections

Rectus abdominis muscle

Skin and subcutaneous tissue

Umbilicus

Anterior layer of rectus sheath

Pyramidalis muscle

Spermatic cord

External oblique muscle (cut away)

Internal oblique muscle

Inguinal ligament

FIGURE 4.2 **The anterolateral abdominal wall.** (A) Superficial and (B) middle layers.

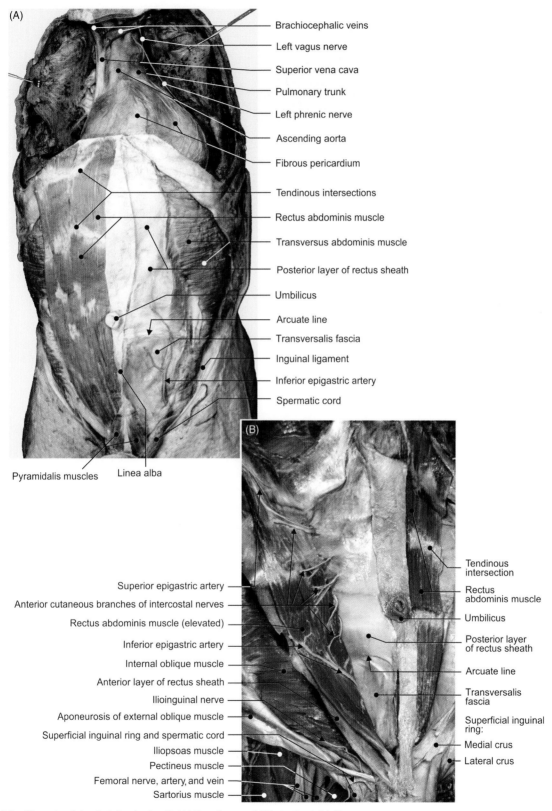

(A)

— Brachiocephalic veins

— Left vagus nerve

— Superior vena cava

— Pulmonary trunk

— Left phrenic nerve

— Ascending aorta

— Fibrous pericardium

— Tendinous intersections

— Rectus abdominis muscle

— Transversus abdominis muscle

— Posterior layer of rectus sheath

— Umbilicus

— Arcuate line

— Transversalis fascia

— Inguinal ligament

— Inferior epigastric artery

— Spermatic cord

Pyramidalis muscles   Linea alba

(B)

Tendinous intersection

Rectus abdominis muscle

Umbilicus

Posterior layer of rectus sheath

Arcuate line

Transversalis fascia

Superficial inguinal ring:

— Medial crus

— Lateral crus

Superior epigastric artery —

Anterior cutaneous branches of intercostal nerves —

Rectus abdominis muscle (elevated) —

Inferior epigastric artery —

Internal oblique muscle —

Anterior layer of rectus sheath —

Ilioinguinal nerve —

Aponeurosis of external oblique muscle —

Superficial inguinal ring and spermatic cord —

Iliopsoas muscle —

Pectineus muscle —

Femoral nerve, artery, and vein —

Sartorius muscle —

FIGURE 4.3   The anterolateral abdominal wall. (A) Deep layer and (B) structural details.

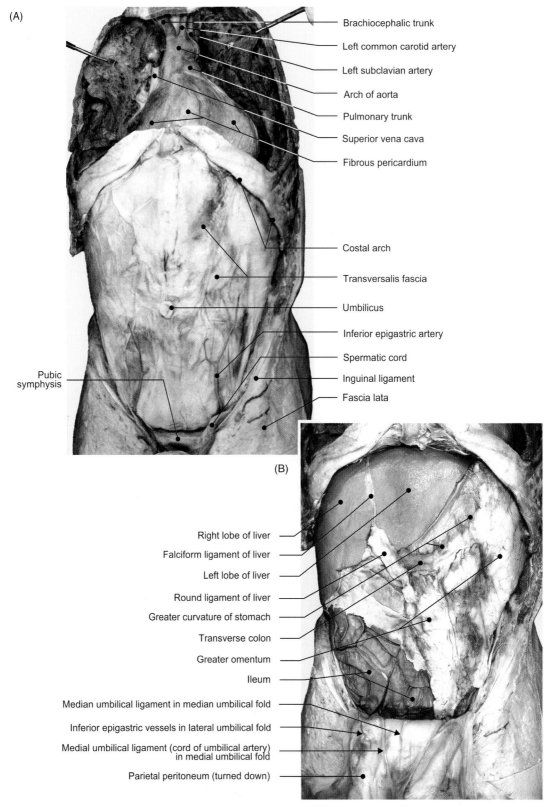

(A)

Brachiocephalic trunk

Left common carotid artery

Left subclavian artery

Arch of aorta

Pulmonary trunk

Superior vena cava

Fibrous pericardium

Costal arch

Transversalis fascia

Umbilicus

Inferior epigastric artery

Spermatic cord

Inguinal ligament

Fascia lata

Pubic symphysis

(B)

Right lobe of liver

Falciform ligament of liver

Left lobe of liver

Round ligament of liver

Greater curvature of stomach

Transverse colon

Greater omentum

Ileum

Median umbilical ligament in median umbilical fold

Inferior epigastric vessels in lateral umbilical fold

Medial umbilical ligament (cord of umbilical artery) in medial umbilical fold

Parietal peritoneum (turned down)

FIGURE 4.4 The anterolateral abdominal wall. (A) Transversalis fascia and (B) exposed anterior aspect of the abdominal content in situ.

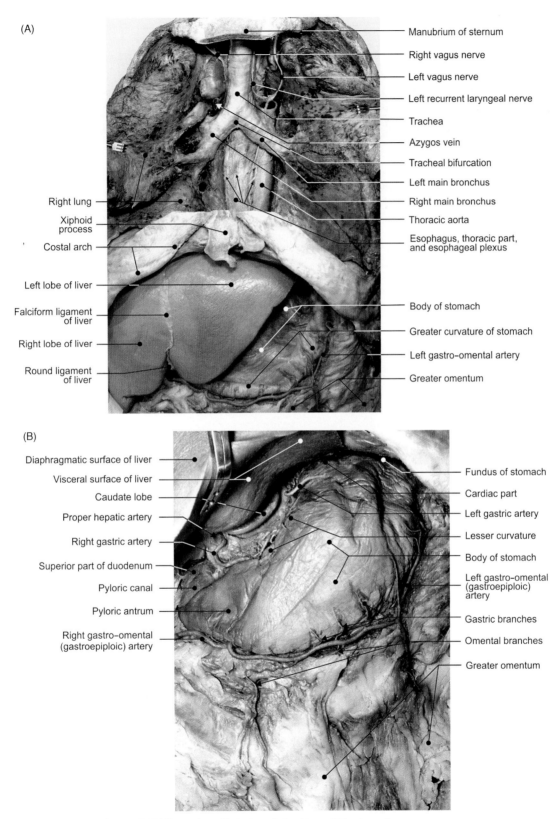

FIGURE 4.5 **The stomach in situ.** (A–B) Different views displaying relationships and blood vessels.

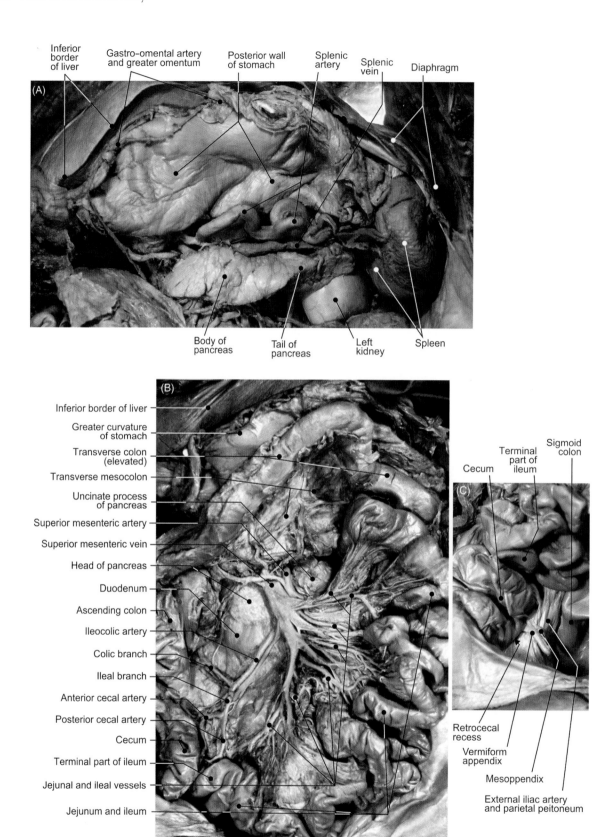

FIGURE 4.6 Abdominal viscera and vascularization. (A) Stomach reflected superiorly: exposed omental bursa. (B) Superior mesenteric artery and vein. (C) Appendix: relationships.

(A)

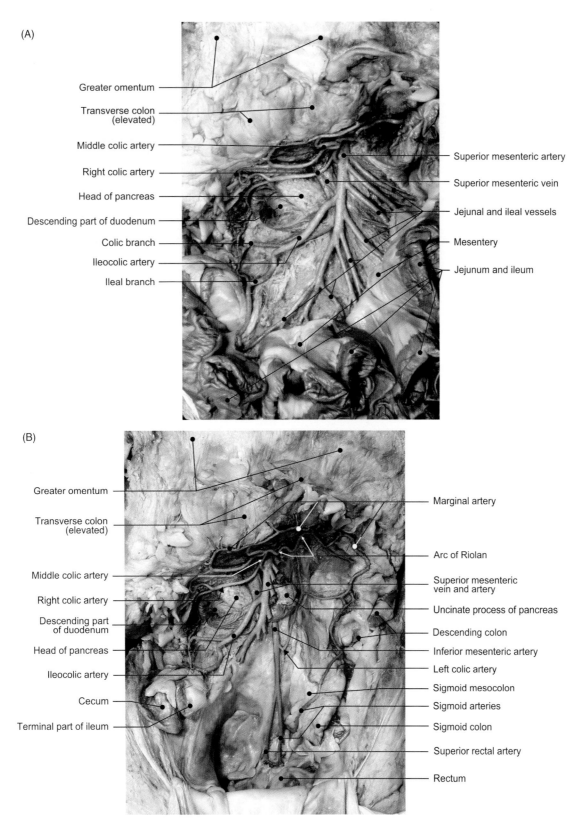

Greater omentum

Transverse colon
(elevated)

Middle colic artery

Right colic artery

Head of pancreas

Descending part of duodenum

Colic branch

Ileocolic artery

Ileal branch

Superior mesenteric artery

Superior mesenteric vein

Jejunal and ileal vessels

Mesentery

Jejunum and ileum

(B)

Greater omentum

Transverse colon
(elevated)

Middle colic artery

Right colic artery

Descending part
of duodenum

Head of pancreas

Ileocolic artery

Cecum

Terminal part of ileum

Marginal artery

Arc of Riolan

Superior mesenteric
vein and artery

Uncinate process of pancreas

Descending colon

Inferior mesenteric artery

Left colic artery

Sigmoid mesocolon

Sigmoid arteries

Sigmoid colon

Superior rectal artery

Rectum

FIGURE 4.7 (A–B) Different views depicting the arterial supply of the small and large intestines.

(A)

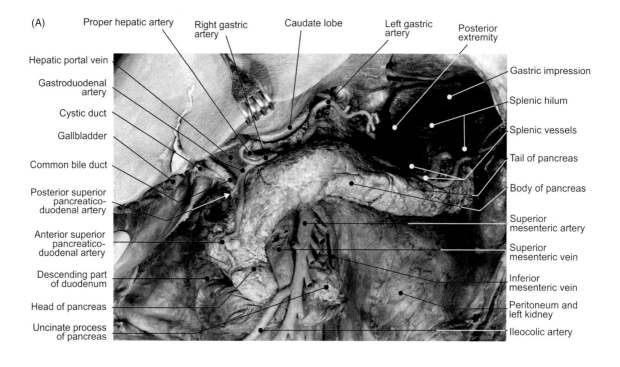

Proper hepatic artery
Right gastric artery
Caudate lobe
Left gastric artery
Posterior extremity

Hepatic portal vein
Gastroduodenal artery
Cystic duct
Gallbladder
Common bile duct
Posterior superior pancreatico-duodenal artery
Anterior superior pancreatico-duodenal artery
Descending part of duodenum
Head of pancreas
Uncinate process of pancreas

Gastric impression
Splenic hilum
Splenic vessels
Tail of pancreas
Body of pancreas
Superior mesenteric artery
Superior mesenteric vein
Inferior mesenteric vein
Peritoneum and left kidney
Ileocolic artery

(B)

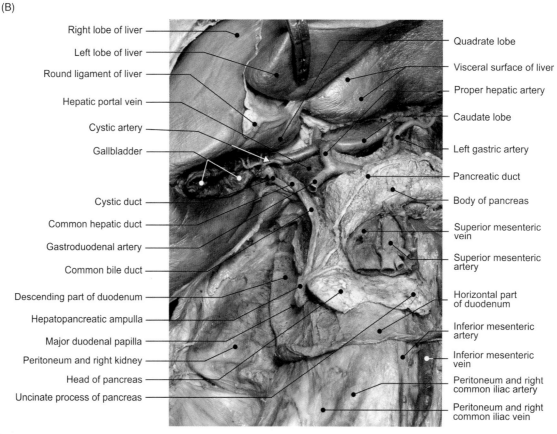

Right lobe of liver
Left lobe of liver
Round ligament of liver
Hepatic portal vein
Cystic artery
Gallbladder
Cystic duct
Common hepatic duct
Gastroduodenal artery
Common bile duct
Descending part of duodenum
Hepatopancreatic ampulla
Major duodenal papilla
Peritoneum and right kidney
Head of pancreas
Uncinate process of pancreas

Quadrate lobe
Visceral surface of liver
Proper hepatic artery
Caudate lobe
Left gastric artery
Pancreatic duct
Body of pancreas
Superior mesenteric vein
Superior mesenteric artery
Horizontal part of duodenum
Inferior mesenteric artery
Inferior mesenteric vein
Peritoneum and right common iliac artery
Peritoneum and right common iliac vein

FIGURE 4.8 (A–B) Different views displaying the liver, pancreas, duodenum, and billiary structures in situ.

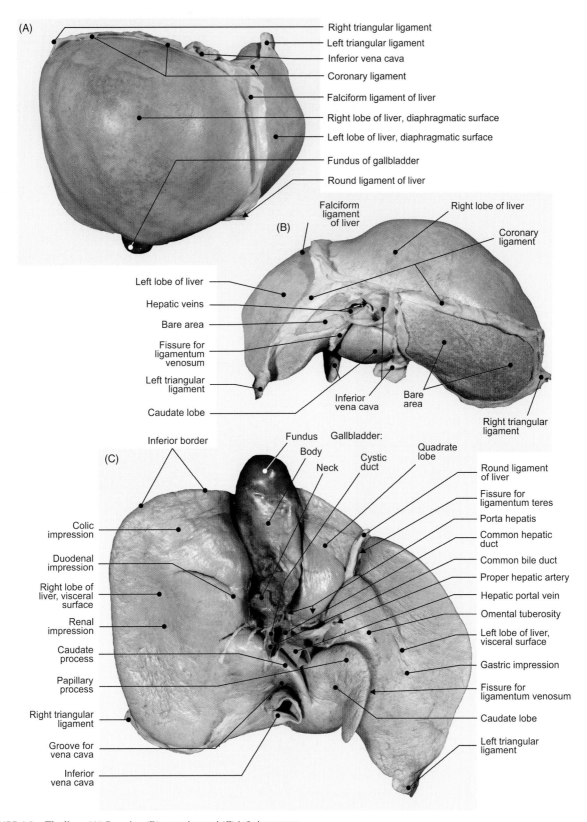

FIGURE 4.9 **The liver.** (A) Superior, (B) posterior, and (C) inferior aspects.

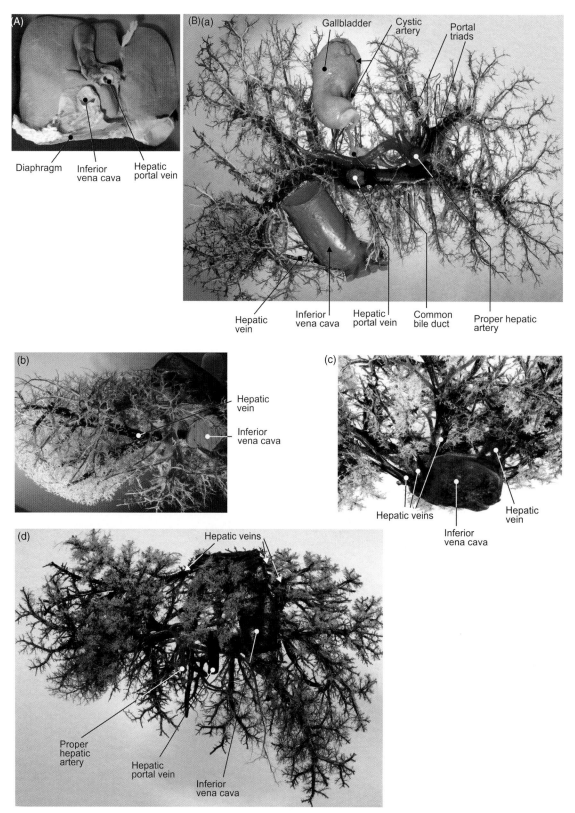

FIGURE 4.10 The liver. (A) Visceral surface of the native liver. (B) Blood circulation (corrosion casts): arborization of hepatic artery proper *(red)* following bile ducts *(yellow)*, and branches of the portal vein *(blue)*. (a) Inferior, (b) superior, (c) detailed superior, and (d) posterior views of the cast.

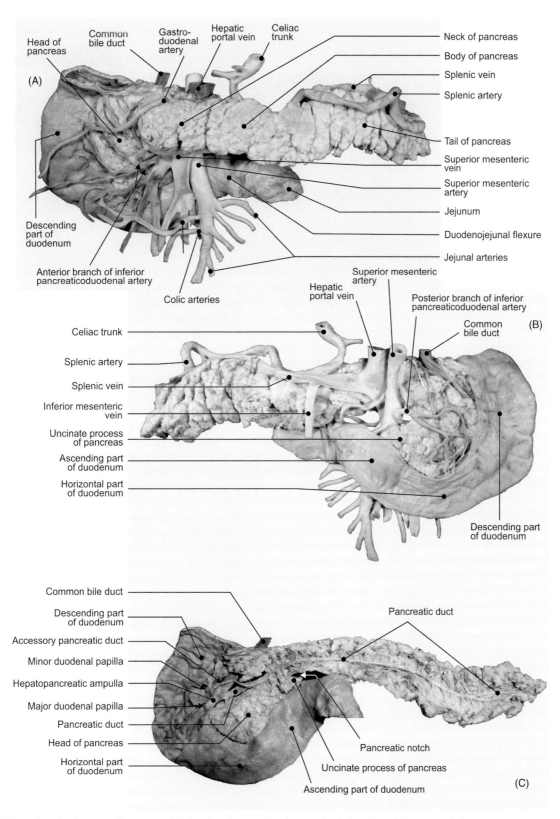

FIGURE 4.11 **The duodenum and pancreas.** (A) Anterior, (B) posterior views, and (C) dissection of the pancreatic duct.

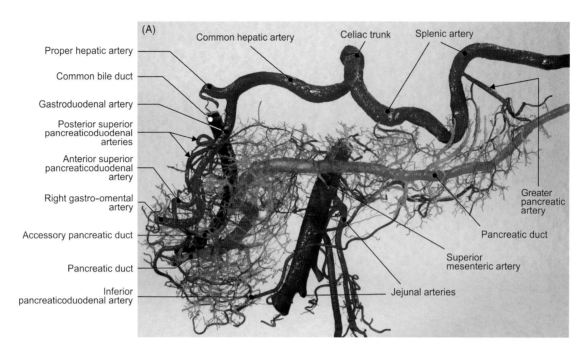

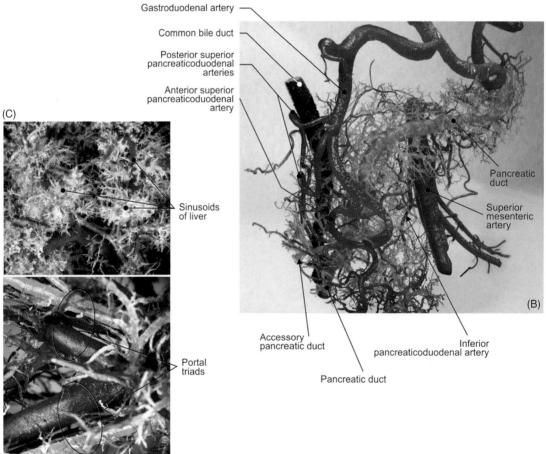

FIGURE 4.12 **Corrosion casts of arteries and excretory pancreatic ducts.** (A) Anterior view and (B) view from the right side. (C) Corrosion casts of the liver.

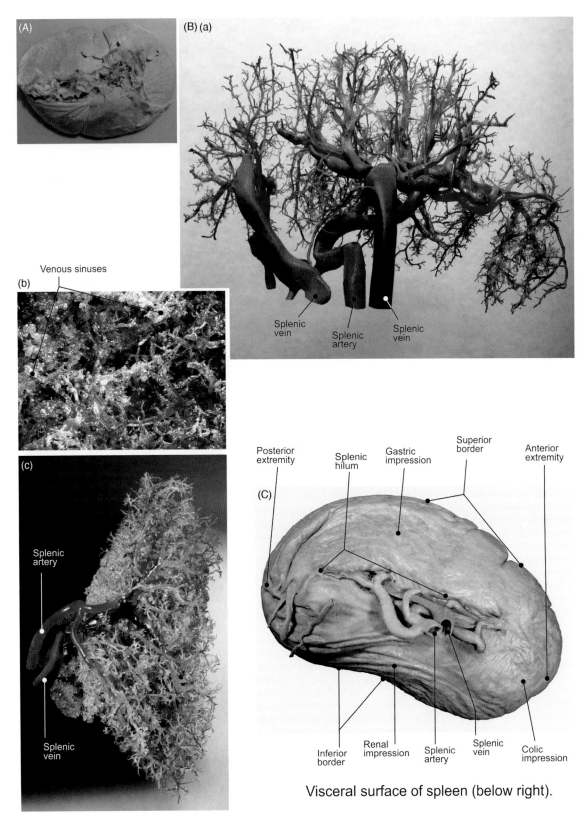

(A)

(B) (a)

Splenic
vein

Splenic
artery

Splenic
vein

Venous sinuses

(b)

(c)

Splenic
artery

Splenic
vein

(C)

Posterior
extremity

Splenic
hilum

Gastric
impression

Superior
border

Anterior
extremity

Inferior
border

Renal
impression

Splenic
artery

Splenic
vein

Colic
impression

Visceral surface of spleen (below right).

FIGURE 4.13   (A) The native spleen: visceral surface. (B) (a–c) Corrosion casts of the splenic artery and vein branches inside of the spleen. (C) Visceral surface of the spleen.

(A)

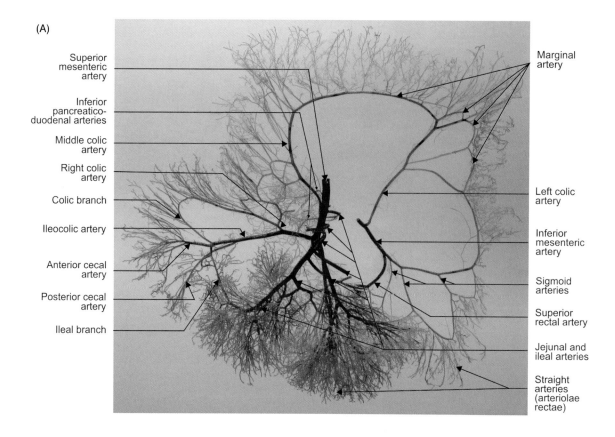

Superior mesenteric artery

Inferior pancreatico-duodenal arteries

Middle colic artery

Right colic artery

Colic branch

Ileocolic artery

Anterior cecal artery

Posterior cecal artery

Ileal branch

Marginal artery

Left colic artery

Inferior mesenteric artery

Sigmoid arteries

Superior rectal artery

Jejunal and ileal arteries

Straight arteries (arteriolae rectae)

(B)

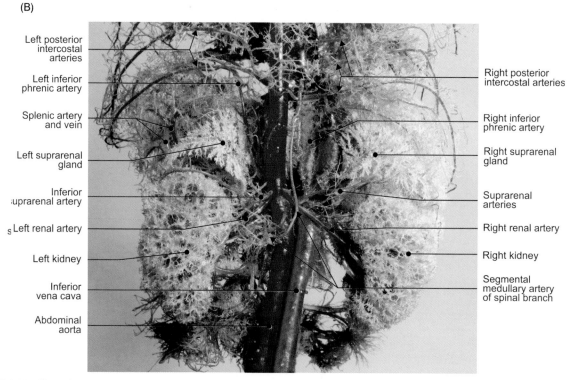

Left posterior intercostal arteries

Left inferior phrenic artery

Splenic artery and vein

Left suprarenal gland

Inferior suprarenal artery

Left renal artery

Left kidney

Inferior vena cava

Abdominal aorta

Right posterior intercostal arteries

Right inferior phrenic artery

Right suprarenal gland

Suprarenal arteries

Right renal artery

Right kidney

Segmental medullary artery of spinal branch

FIGURE 4.14 Corrosion casts of the fetus. (A) Anterior aspect of the intestinal arteries and (B) posterior aspect of the visceral abdominal blood vessels.

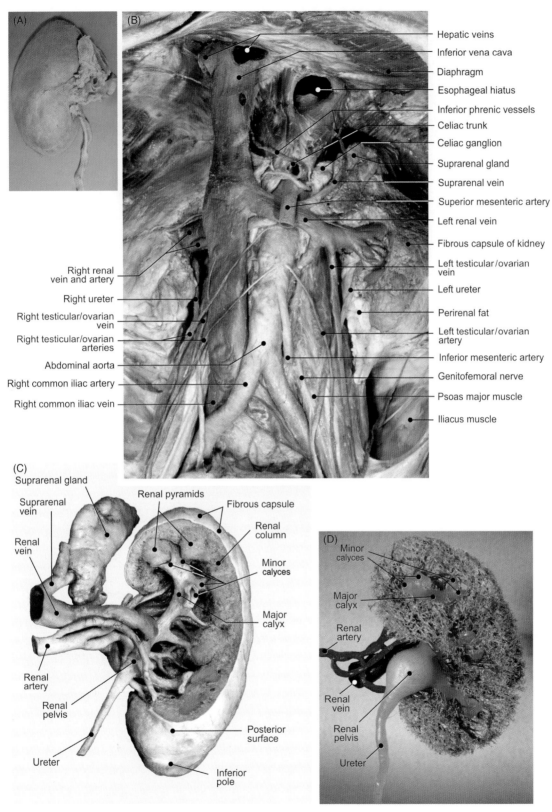

(A)

(B)

Hepatic veins
Inferior vena cava
Diaphragm
Esophageal hiatus
Inferior phrenic vessels
Celiac trunk
Celiac ganglion
Suprarenal gland
Suprarenal vein
Superior mesenteric artery
Left renal vein
Fibrous capsule of kidney
Left testicular/ovarian vein
Left ureter
Perirenal fat
Left testicular/ovarian artery
Inferior mesenteric artery
Genitofemoral nerve
Psoas major muscle
Iliacus muscle

Right renal vein and artery
Right ureter
Right testicular/ovarian vein
Right testicular/ovarian arteries
Abdominal aorta
Right common iliac artery
Right common iliac vein

(C)
Suprarenal gland
Suprarenal vein
Renal vein
Renal artery
Renal pelvis
Ureter

Renal pyramids
Fibrous capsule
Renal column
Minor calyces
Major calyx
Posterior surface
Inferior pole

(D)
Minor calyces
Major calyx
Renal artery
Renal vein
Renal pelvis
Ureter

FIGURE 4.15 (A) The native kidney: anterior surface. (B) Kidneys in situ: anterior aspect. Posterior aspect of the right kidney: (C) dissection and (D) corrosion cast.

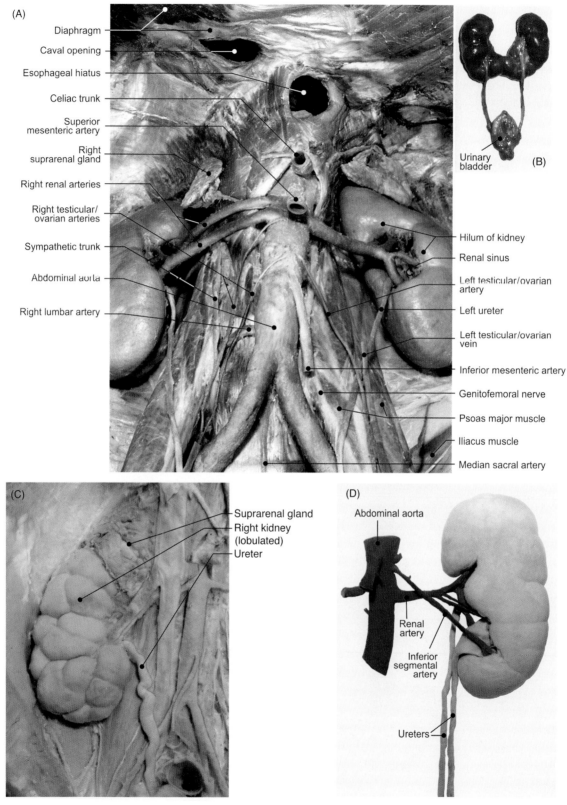

FIGURE 4.16 (A) Anterior aspect of the kidneys after the removal of veins. Newborn: (B) arcuate kidney and (C) right suprarenal gland, lobulated kidney, and ureter in situ. (D) Double ureter and accessory inferior segmental artery.

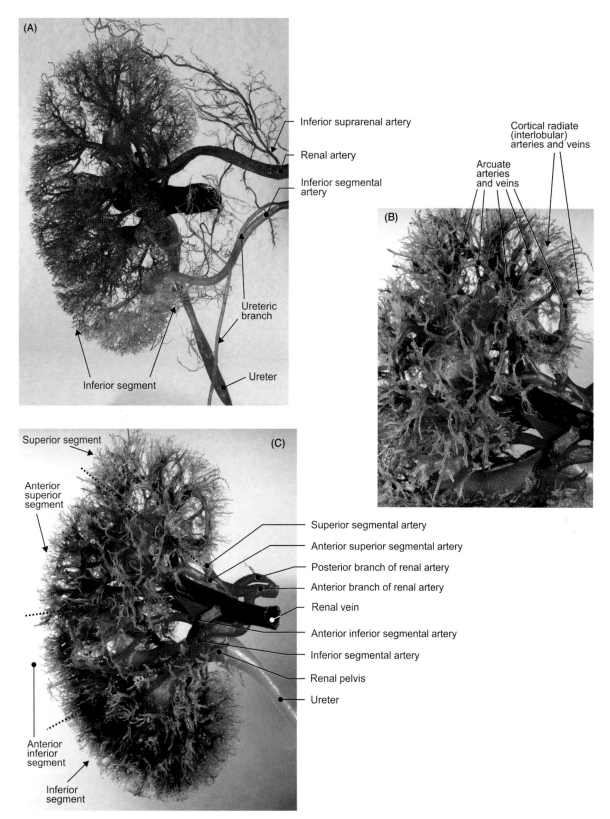

FIGURE 4.17 (A–C) Different views of the anterior aspect of the right kidney (corrosion casts).

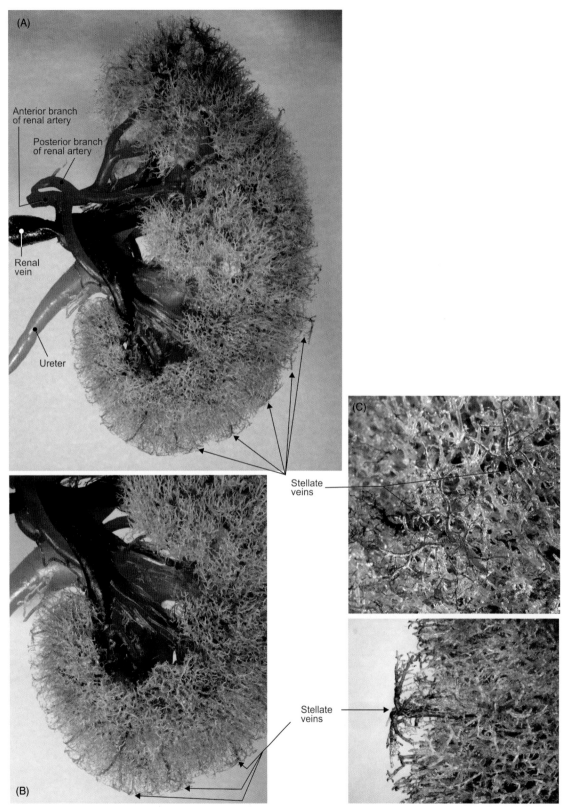

FIGURE 4.18 (A–C) Different views of the posterior aspect of the right kidney (corrosion casts).

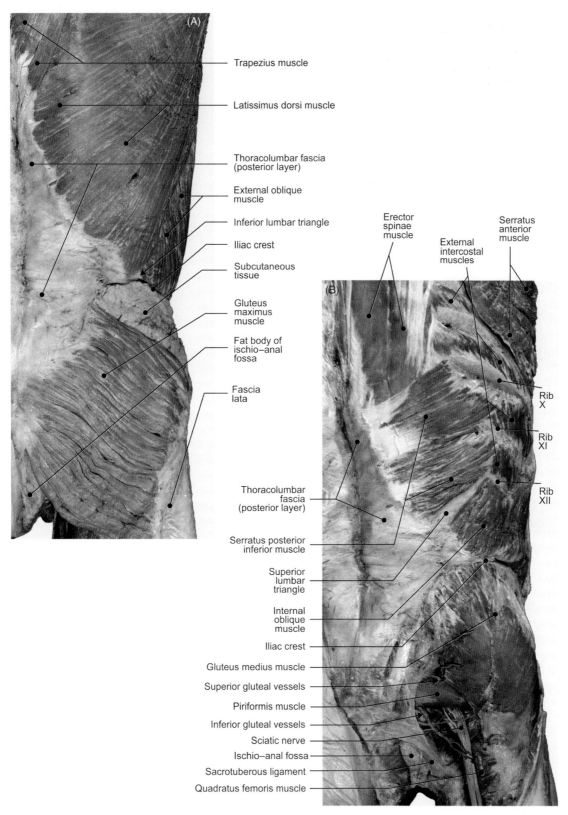

FIGURE 4.19 The posterior aspect of the abdominal wall and gluteal region. (A) Superficial and (B) middle layers.

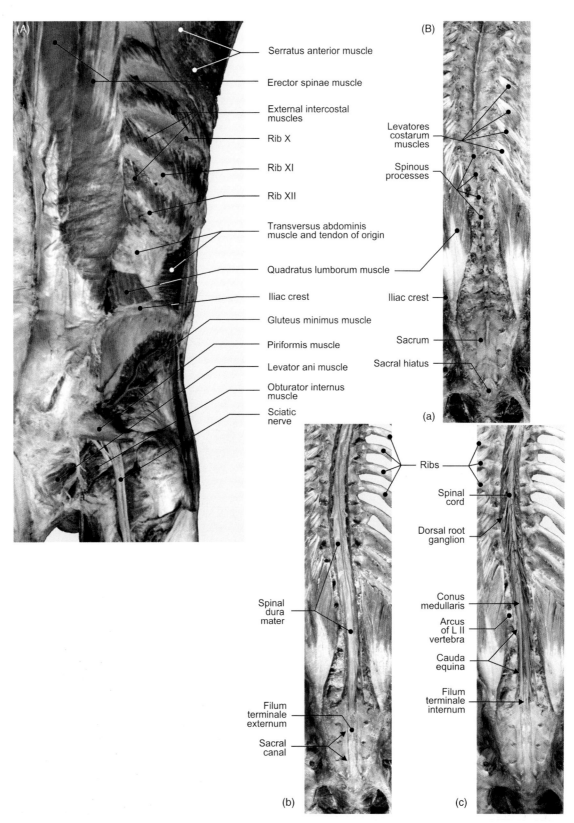

(A)

Serratus anterior muscle

Erector spinae muscle

External intercostal
muscles

Rib X

Rib XI

Rib XII

Transversus abdominis
muscle and tendon of origin

Quadratus lumborum muscle

Iliac crest

Gluteus minimus muscle

Piriformis muscle

Levator ani muscle

Obturator internus
muscle

Sciatic
nerve

(B)

Levatores
costarum
muscles

Spinous
processes

Iliac crest

Sacrum

Sacral hiatus

(a)

Ribs

Spinal
cord

Dorsal root
ganglion

Spinal
dura
mater

Conus
medullaris

Arcus
of L II
vertebra

Cauda
equina

Filum
terminale
externum

Filum
terminale
internum

Sacral
canal

(b)

(c)

FIGURE 4.20   (A) Posterior aspect of the abdominal wall and gluteal region: deep layer. (B) Posterior views of the vertebral column: (a) spinal processes; (b) dura mater, after the spinal processes were removed; and (c) spinal cord, after dura mater was removed.

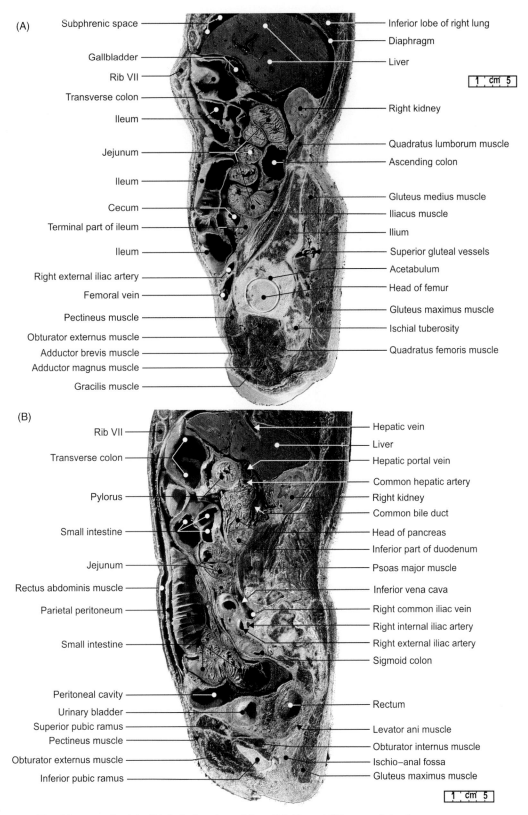

(A)

Subphrenic space

Gallbladder

Rib VII

Transverse colon

Ileum

Jejunum

Ileum

Cecum

Terminal part of ileum

Ileum

Right external iliac artery

Femoral vein

Pectineus muscle

Obturator externus muscle

Adductor brevis muscle

Adductor magnus muscle

Gracilis muscle

Inferior lobe of right lung

Diaphragm

Liver

1 cm 5

Right kidney

Quadratus lumborum muscle

Ascending colon

Gluteus medius muscle

Iliacus muscle

Ilium

Superior gluteal vessels

Acetabulum

Head of femur

Gluteus maximus muscle

Ischial tuberosity

Quadratus femoris muscle

(B)

Rib VII

Transverse colon

Pylorus

Small intestine

Jejunum

Rectus abdominis muscle

Parietal peritoneum

Small intestine

Peritoneal cavity

Urinary bladder

Superior pubic ramus

Pectineus muscle

Obturator externus muscle

Inferior pubic ramus

Hepatic vein

Liver

Hepatic portal vein

Common hepatic artery

Right kidney

Common bile duct

Head of pancreas

Inferior part of duodenum

Psoas major muscle

Inferior vena cava

Right common iliac vein

Right internal iliac artery

Right external iliac artery

Sigmoid colon

Rectum

Levator ani muscle

Obturator internus muscle

Ischio–anal fossa

Gluteus maximus muscle

1 cm 5

FIGURE 4.21 **The abdomen and pelvis.** (A) Sagittal sections of the gallbladder and (B) pancreatic head.

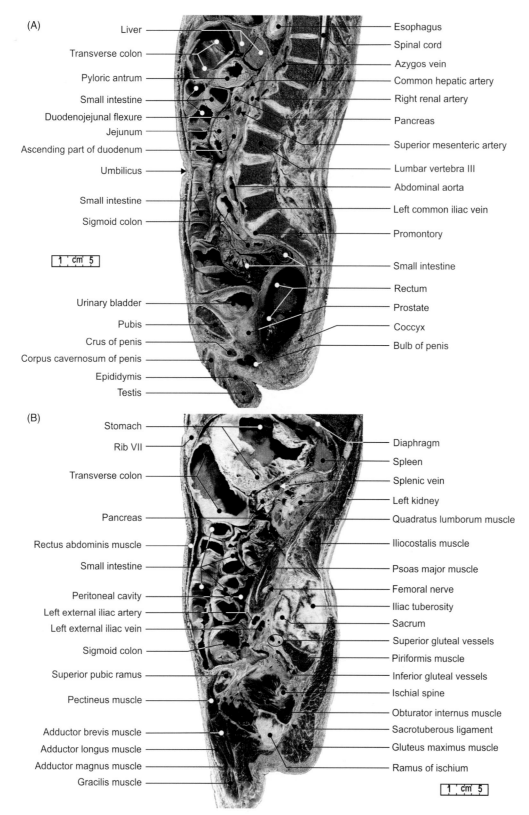

(A)

Liver

Transverse colon

Pyloric antrum

Small intestine

Duodenojejunal flexure

Jejunum

Ascending part of duodenum

Umbilicus

Small intestine

Sigmoid colon

1 cm 5

Urinary bladder

Pubis

Crus of penis

Corpus cavernosum of penis

Epididymis

Testis

Esophagus

Spinal cord

Azygos vein

Common hepatic artery

Right renal artery

Pancreas

Superior mesenteric artery

Lumbar vertebra III

Abdominal aorta

Left common iliac vein

Promontory

Small intestine

Rectum

Prostate

Coccyx

Bulb of penis

(B)

Stomach

Rib VII

Transverse colon

Pancreas

Rectus abdominis muscle

Small intestine

Peritoneal cavity

Left external iliac artery

Left external iliac vein

Sigmoid colon

Superior pubic ramus

Pectineus muscle

Adductor brevis muscle

Adductor longus muscle

Adductor magnus muscle

Gracilis muscle

Diaphragm

Spleen

Splenic vein

Left kidney

Quadratus lumborum muscle

Iliocostalis muscle

Psoas major muscle

Femoral nerve

Iliac tuberosity

Sacrum

Superior gluteal vessels

Piriformis muscle

Inferior gluteal vessels

Ischial spine

Obturator internus muscle

Sacrotuberous ligament

Gluteus maximus muscle

Ramus of ischium

1 cm 5

FIGURE 4.22 **The abdomen and pelvis.** (A) Sagittal sections of the spinal canal and cord and (B) the left kidney.

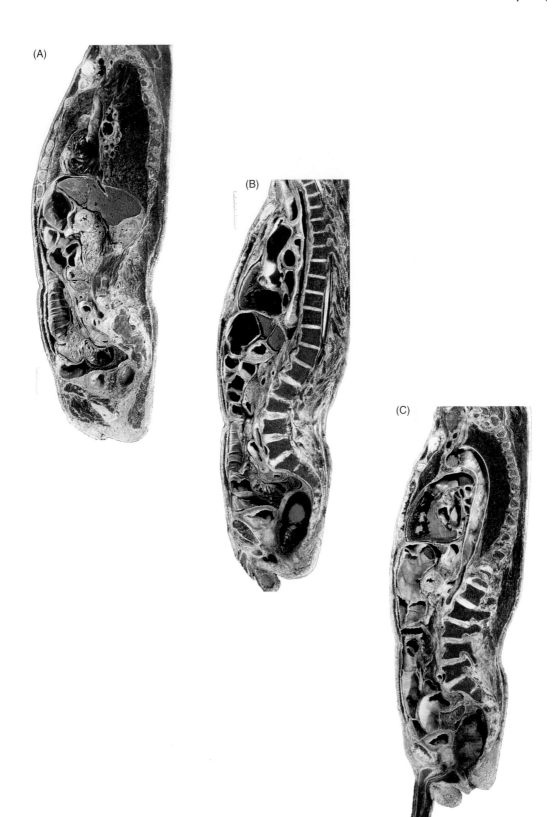

FIGURE 4.23   Sections of the human body (refer to Figs. 3.25A,B, 3.26A, 4.21B, and 4.22A). (A) Right sagittal section of the pancreatic head and right kidney. (B) Midsagittal and (C) left sagittal sections of the descending aorta.

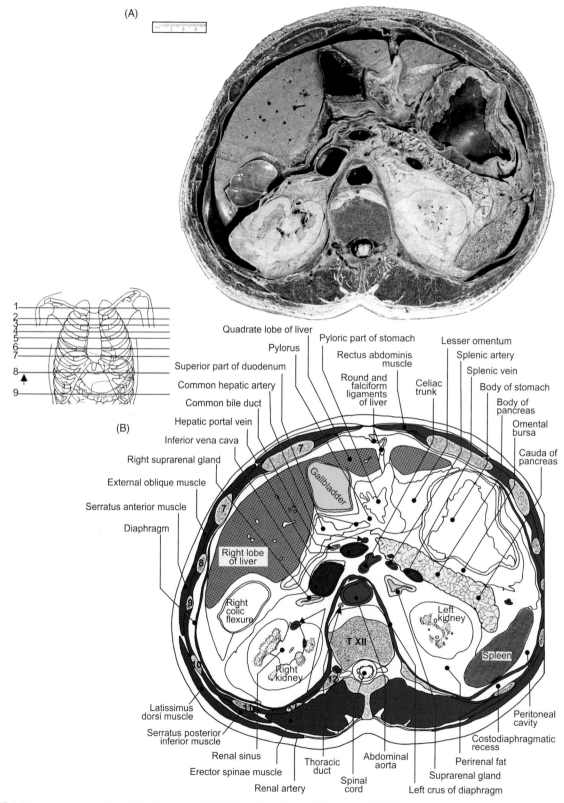

FIGURE 4.24 Transverse section of the abdomen at the T12 vertebral level. (A) Section and (B) illustration.

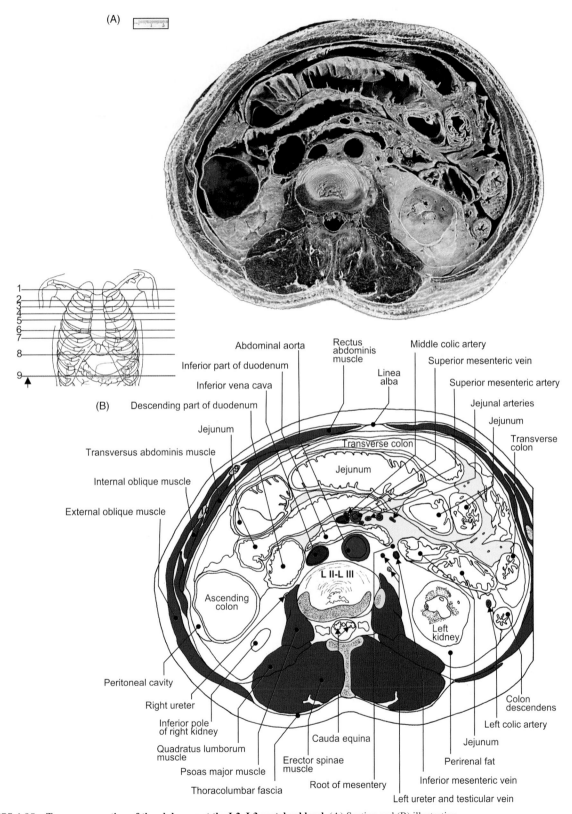

FIGURE 4.25   Transverse section of the abdomen at the L2–L3 vertebral level. (A) Section and (B) illustration.

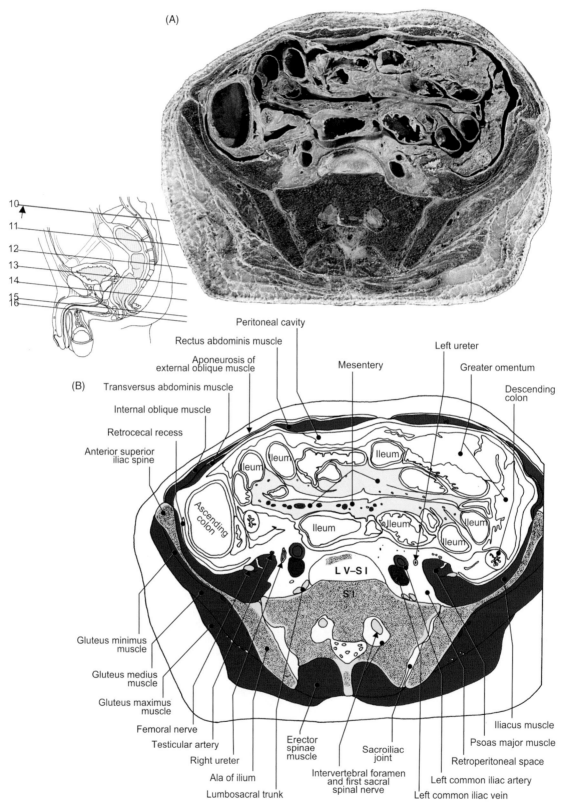

(A)

10
11
12
13
14
15
16

(B)

Peritoneal cavity

Rectus abdominis muscle

Mesentery

Left ureter

Greater omentum

Aponeurosis of
external oblique muscle

Transversus abdominis muscle

Descending
colon

Internal oblique muscle

Retrocecal recess

Anterior superior
iliac spine

Ileum

Ileum

Ileum

Ascending
colon

Ileum

Ileum

Ileum

Ileum

Ileum

L V–S I

S1

Gluteus minimus
muscle

Gluteus medius
muscle

Gluteus maximus
muscle

Femoral nerve

Iliacus muscle

Testicular artery

Erector
spinae
muscle

Psoas major muscle

Sacroiliac
joint

Retroperitoneal space

Right ureter

Ala of ilium

Intervertebral foramen
and first sacral
spinal nerve

Left common iliac artery

Left common iliac vein

Lumbosacral trunk

FIGURE 4.26 **Transverse section of the abdomen at the L5–S1 vertebral level.** (A) Section and (B) illustration.

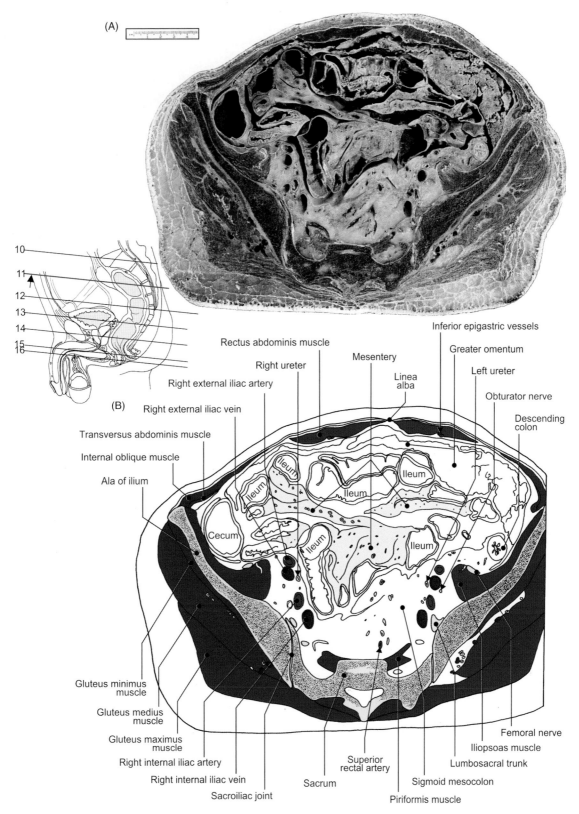

FIGURE 4.27 **Transverse section of the abdomen and pelvis at the midpelvic level.** (A) Section and (B) illustration.

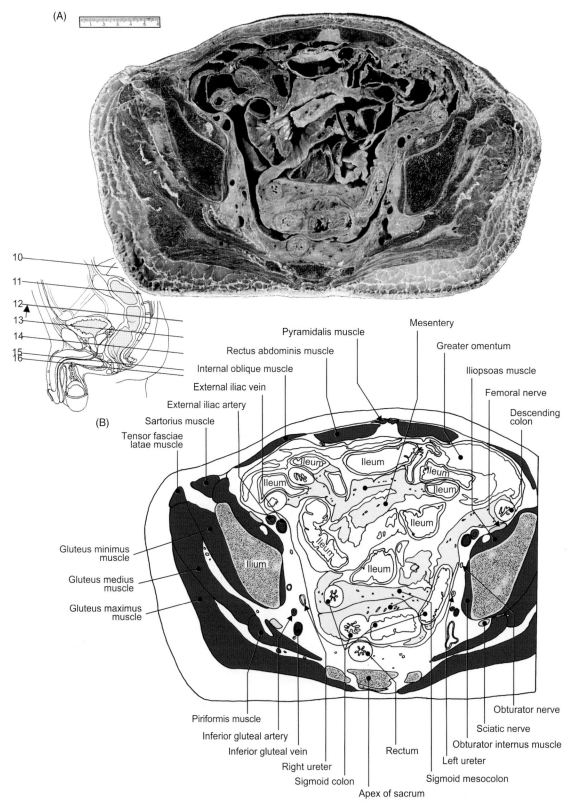

(A)

(B)

10
11
12
13
14
15
16

Pyramidalis muscle

Mesentery

Rectus abdominis muscle

Greater omentum

Internal oblique muscle

Iliopsoas muscle

External iliac vein

Femoral nerve

External iliac artery

Descending colon

Sartorius muscle

Tensor fasciae latae muscle

Ileum

Ileum

Ileum

Ileum

Ileum

Ileum

Gluteus minimus muscle

Ilium

Ileum

Gluteus medius muscle

Gluteus maximus muscle

Piriformis muscle

Inferior gluteal artery

Inferior gluteal vein

Right ureter

Sigmoid colon

Apex of sacrum

Rectum

Left ureter

Sigmoid mesocolon

Obturator internus muscle

Sciatic nerve

Obturator nerve

FIGURE 4.28 Transverse section of the abdomen and pelvis at the sacral tip level. (A) Section and (B) illustration.

Chapter 5

Pelvis and Perineum with 5–6-Month-Old Fetal Specimens

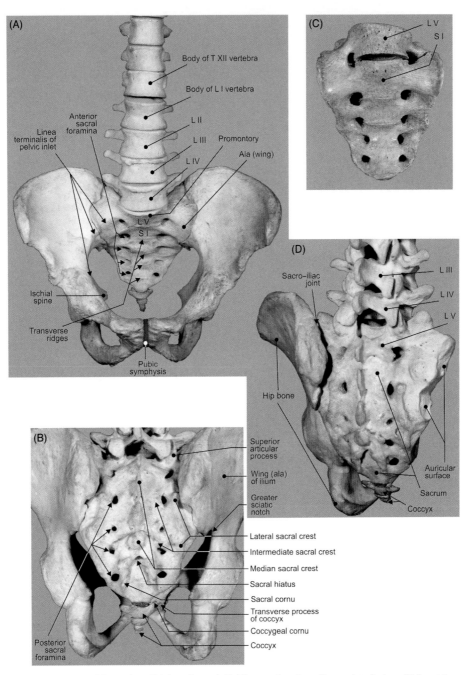

FIGURE 5.1 The sacrum and coccyx. L5 vertebra: (A) Anterior and (B–D) posterior views. Incomplete fusion of L5 vertebra with sacrum: anterior view and total sacralization of L5 vertebra.

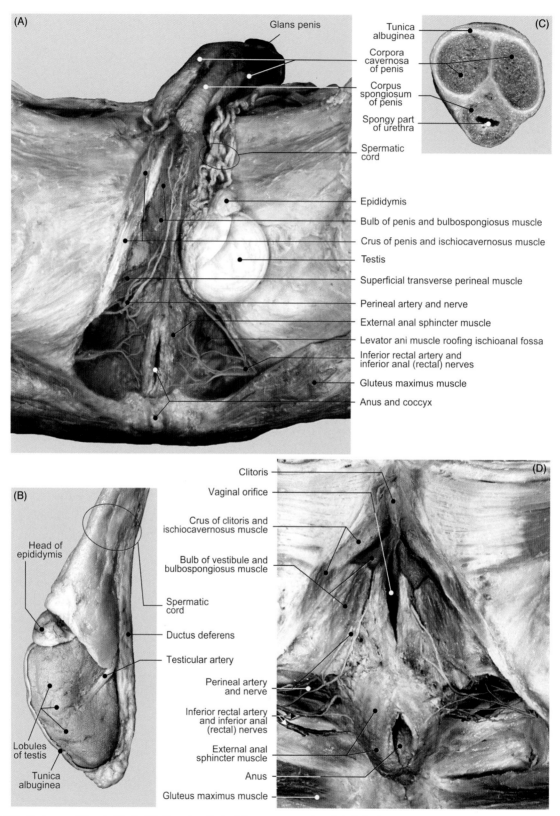

FIGURE 5.2 Perineum: (A) male. Testis: (B) dissection. Penis: (C) transverse section of shaft. Perineum: (D) female.

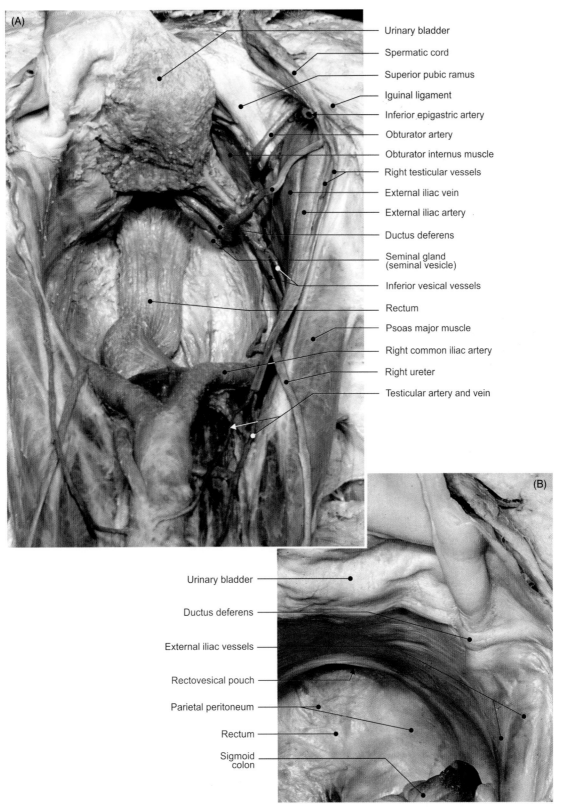

Urinary bladder
Spermatic cord
Superior pubic ramus
Iguinal ligament
Inferior epigastric artery
Obturator artery
Obturator internus muscle
Right testicular vessels
External iliac vein
External iliac artery
Ductus deferens
Seminal gland (seminal vesicle)
Inferior vesical vessels
Rectum
Psoas major muscle
Right common iliac artery
Right ureter
Testicular artery and vein

Urinary bladder
Ductus deferens
External iliac vessels
Rectovesical pouch
Parietal peritoneum
Rectum
Sigmoid colon

FIGURE 5.3 **The male pelvis.** View from above: (A) organs with peritoneum removed and (B) organs covered by peritoneum.

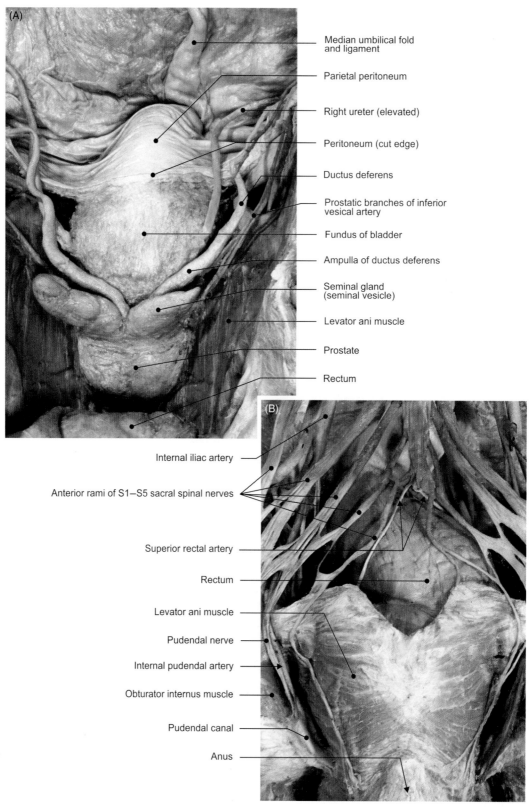

(A)

Median umbilical fold and ligament

Parietal peritoneum

Right ureter (elevated)

Peritoneum (cut edge)

Ductus deferens

Prostatic branches of inferior vesical artery

Fundus of bladder

Ampulla of ductus deferens

Seminal gland (seminal vesicle)

Levator ani muscle

Prostate

Rectum

Internal iliac artery

Anterior rami of S1—S5 sacral spinal nerves

Superior rectal artery

Rectum

Levator ani muscle

Pudendal nerve

Internal pudendal artery

Obturator internus muscle

Pudendal canal

Anus

(B)

FIGURE 5.4 **The prostate and urinary bladder.** (A) View from behind. Levator ani muscle and branches of sacral plexus: (B) view from behind.

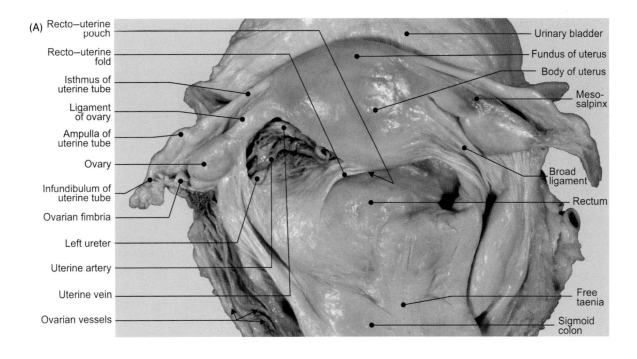

(A) Recto–uterine pouch
Recto–uterine fold
Isthmus of uterine tube
Ligament of ovary
Ampulla of uterine tube
Ovary
Infundibulum of uterine tube
Ovarian fimbria
Left ureter
Uterine artery
Uterine vein
Ovarian vessels

Urinary bladder
Fundus of uterus
Body of uterus
Meso-salpinx
Broad ligament
Rectum
Free taenia
Sigmoid colon

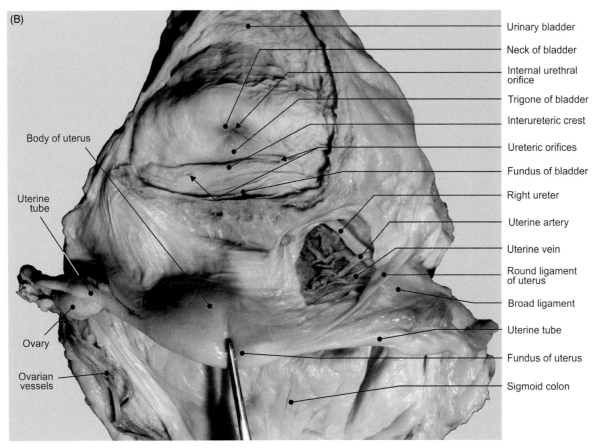

(B)
Body of uterus
Uterine tube
Ovary
Ovarian vessels

Urinary bladder
Neck of bladder
Internal urethral orifice
Trigone of bladder
Interureteric crest
Ureteric orifices
Fundus of bladder
Right ureter
Uterine artery
Uterine vein
Round ligament of uterus
Broad ligament
Uterine tube
Fundus of uterus
Sigmoid colon

FIGURE 5.5   The female internal genital organs viewed from above. (A) Open broad ligament. Female internal genital organs viewed from above: (B) uterus is reflected and broad ligament and urinary bladder opened.

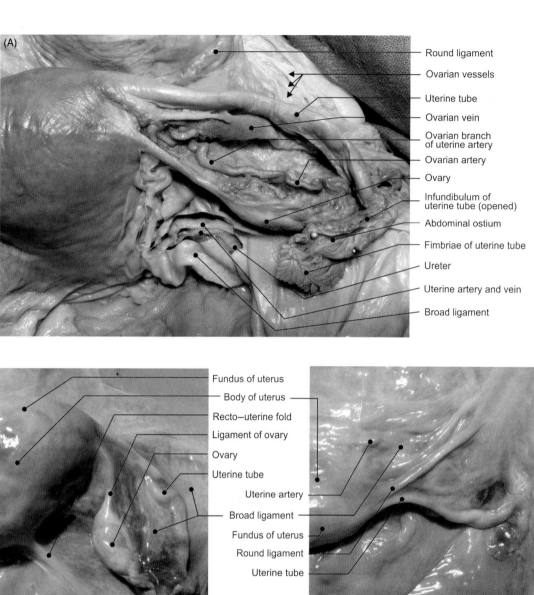

Round ligament
Ovarian vessels
Uterine tube
Ovarian vein
Ovarian branch of uterine artery
Ovarian artery
Ovary
Infundibulum of uterine tube (opened)
Abdominal ostium
Fimbriae of uterine tube
Ureter
Uterine artery and vein
Broad ligament

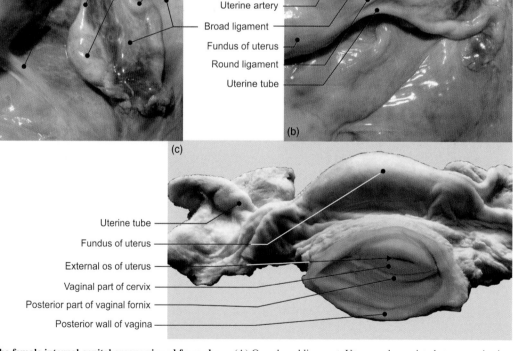

Fundus of uterus
Body of uterus
Recto–uterine fold
Ligament of ovary
Ovary
Uterine tube
Uterine artery
Broad ligament
Fundus of uterus
Round ligament
Uterine tube

Uterine tube
Fundus of uterus
External os of uterus
Vaginal part of cervix
Posterior part of vaginal fornix
Posterior wall of vagina

FIGURE 5.6   The female internal genital organs viewed from above. (A) Open broad ligament. Uterus and associated structures in situ: view from above on (a) posterior and (b) anterior surfaces of native samples, and (c) view from the front.

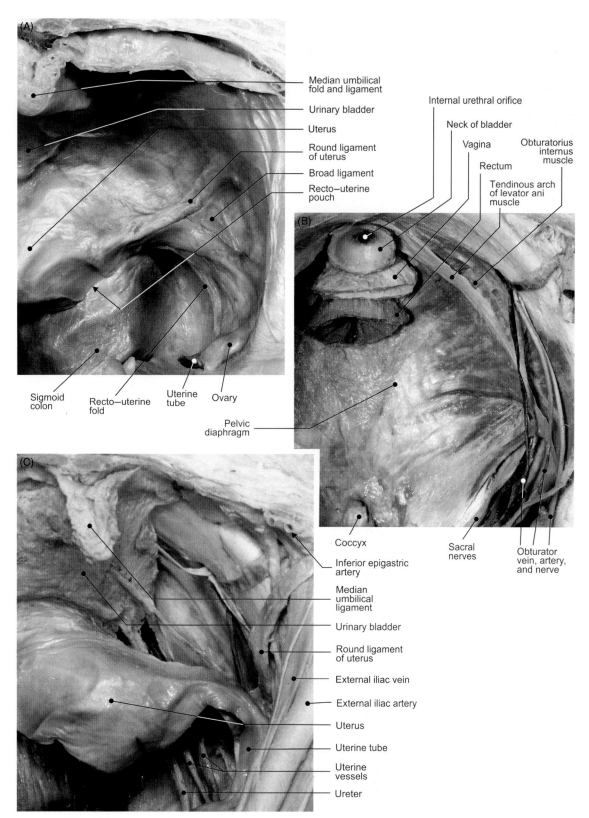

FIGURE 5.7   The female pelvic organs covered by peritoneum. (A) View from above. Female pelvic diaphragm: (B) view from above. Female pelvic organs with peritoneum removed: (C) view from above.

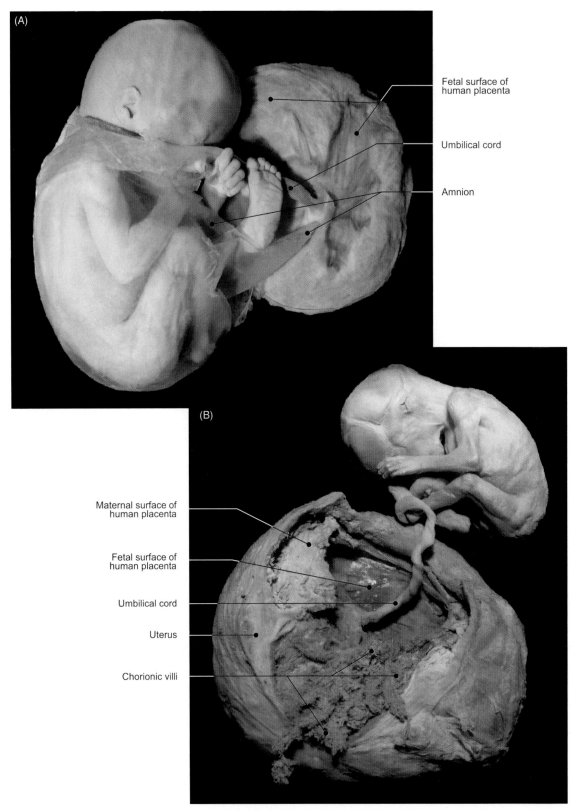

FIGURE 5.8   Five- to six-month-old fetal specimens. (A–B) Different specimens demonstrating the fetus–uterus relationship.

(A)

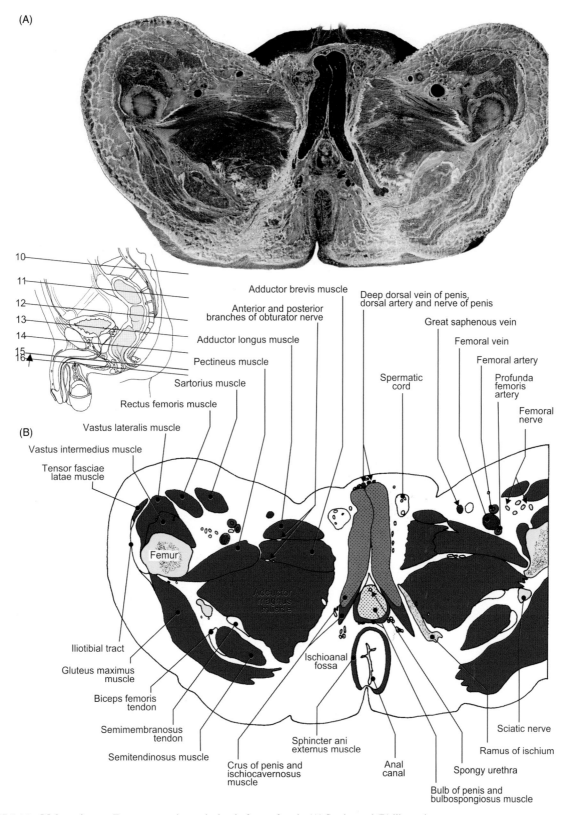

(B)

10
11
12
13
14
15
16

Adductor brevis muscle

Anterior and posterior
branches of obturator nerve

Adductor longus muscle

Pectineus muscle

Sartorius muscle

Rectus femoris muscle

Vastus lateralis muscle

Vastus intermedius muscle

Tensor fasciae
latae muscle

Deep dorsal vein of penis,
dorsal artery and nerve of penis

Great saphenous vein

Femoral vein

Femoral artery

Profunda
femoris
artery

Femoral
nerve

Spermatic
cord

Femur

Adductor
magnus
muscle

Iliotibial tract

Gluteus maximus
muscle

Biceps femoris
tendon

Semimembranosus
tendon

Semitendinosus muscle

Crus of penis and
ischiocavernosus
muscle

Ischioanal
fossa

Sphincter ani
externus muscle

Anal
canal

Bulb of penis and
bulbospongiosus muscle

Spongy urethra

Ramus of ischium

Sciatic nerve

FIGURE 5.11 **Male perineum.** Transverse section at the level of root of penis. (A) Section and (B) illustration.

(A)

(B)

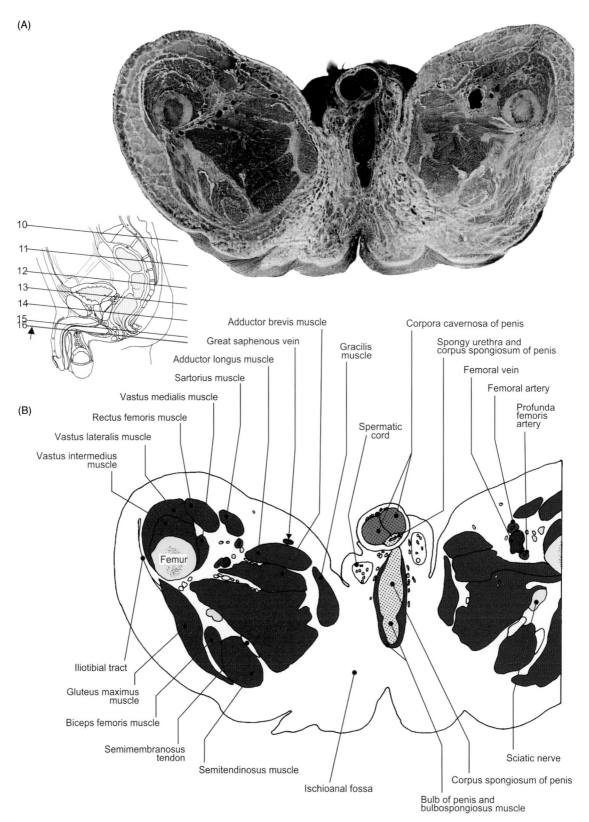

10
11
12
13
14
15
16

Adductor brevis muscle

Great saphenous vein

Adductor longus muscle

Sartorius muscle

Vastus medialis muscle

Rectus femoris muscle

Vastus lateralis muscle

Vastus intermedius
muscle

Gracilis
muscle

Spermatic
cord

Corpora cavernosa of penis

Spongy urethra and
corpus spongiosum of penis

Femoral vein

Femoral artery

Profunda
femoris
artery

Femur

Iliotibial tract

Gluteus maximus
muscle

Biceps femoris muscle

Semimembranosus
tendon

Semitendinosus muscle

Ischioanal fossa

Bulb of penis and
bulbospongiosus muscle

Corpus spongiosum of penis

Sciatic nerve

FIGURE 5.12   The transverse section of the perineum and upper region of thigh. (A) Section and (B) illustration.

Chapter 6

Head and Neck Regions and Vascularization

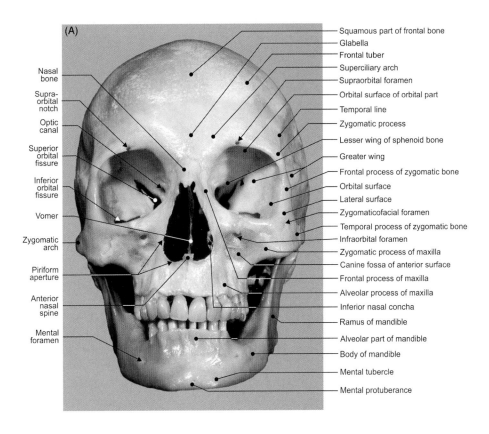

Nasal bone
Supra-orbital notch
Optic canal
Superior orbital fissure
Inferior orbital fissure
Vomer
Zygomatic arch
Piriform aperture
Anterior nasal spine
Mental foramen

Squamous part of frontal bone
Glabella
Frontal tuber
Superciliary arch
Supraorbital foramen
Orbital surface of orbital part
Temporal line
Zygomatic process
Lesser wing of sphenoid bone
Greater wing
Frontal process of zygomatic bone
Orbital surface
Lateral surface
Zygomaticofacial foramen
Temporal process of zygomatic bone
Infraorbital foramen
Zygomatic process of maxilla
Canine fossa of anterior surface
Frontal process of maxilla
Alveolar process of maxilla
Inferior nasal concha
Ramus of mandible
Alveolar part of mandible
Body of mandible
Mental tubercle
Mental protuberance

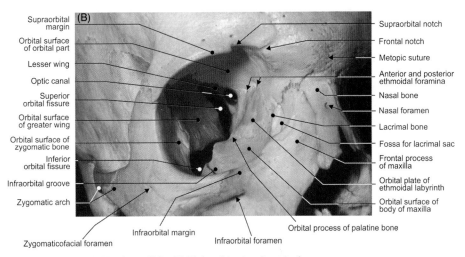

Supraorbital margin
Orbital surface of orbital part
Lesser wing
Optic canal
Superior orbital fissure
Orbital surface of greater wing
Orbital surface of zygomatic bone
Inferior orbital fissure
Infraorbital groove
Zygomatic arch

Supraorbital notch
Frontal notch
Metopic suture
Anterior and posterior ethmoidal foramina
Nasal bone
Nasal foramen
Lacrimal bone
Fossa for lacrimal sac
Frontal process of maxilla
Orbital plate of ethmoidal labyrinth
Orbital surface of body of maxilla

Zygomaticofacial foramen
Infraorbital margin
Infraorbital foramen
Orbital process of palatine bone

FIGURE 6.1 (A) The anterior surface of skull and mandible. (B) Right orbit: view from the front.

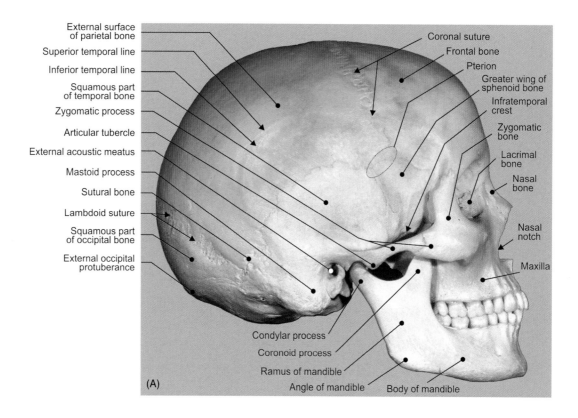

External surface of parietal bone
Superior temporal line
Inferior temporal line
Squamous part of temporal bone
Zygomatic process
Articular tubercle
External acoustic meatus
Mastoid process
Sutural bone
Lambdoid suture
Squamous part of occipital bone
External occipital protuberance

Coronal suture
Frontal bone
Pterion
Greater wing of sphenoid bone
Infratemporal crest
Zygomatic bone
Lacrimal bone
Nasal bone
Nasal notch
Maxilla

Condylar process
Coronoid process
Ramus of mandible
Angle of mandible
Body of mandible

(A)

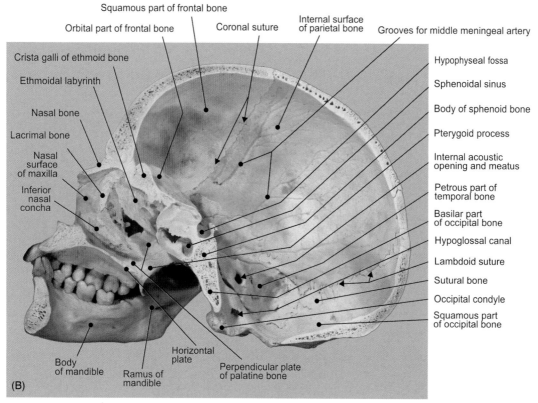

Squamous part of frontal bone
Orbital part of frontal bone
Coronal suture
Internal surface of parietal bone
Grooves for middle meningeal artery

Crista galli of ethmoid bone
Ethmoidal labyrinth
Nasal bone
Lacrimal bone
Nasal surface of maxilla
Inferior nasal concha

Hypophyseal fossa
Sphenoidal sinus
Body of sphenoid bone
Pterygoid process
Internal acoustic opening and meatus
Petrous part of temporal bone
Basilar part of occipital bone
Hypoglossal canal
Lambdoid suture
Sutural bone
Occipital condyle
Squamous part of occipital bone

Body of mandible
Ramus of mandible
Horizontal plate
Perpendicular plate of palatine bone

(B)

FIGURE 6.2 (A) The lateral surface of cranium and mandible. (B) Midsagittal section of cranium and mandible, right half: inside view.

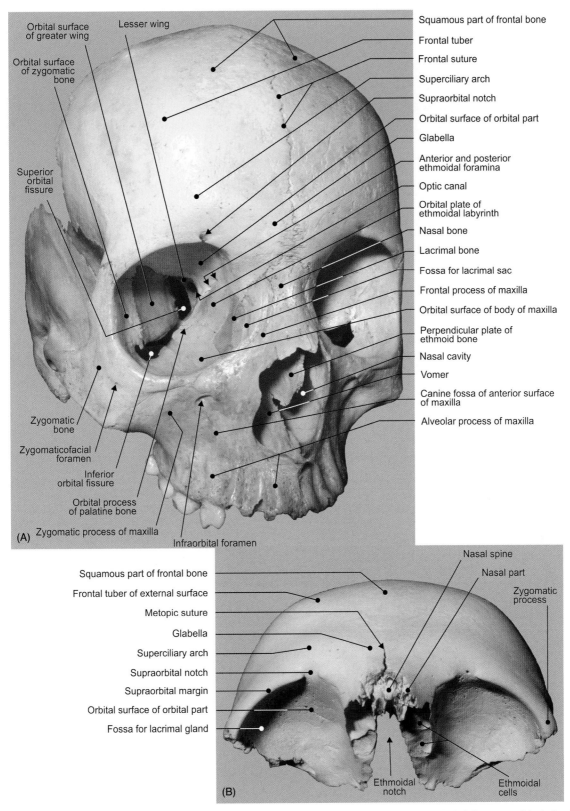

FIGURE 6.3 (A) The anterior aspect of cranial segment: frontal bone divided by persistent frontal suture. (B) The inferior aspect of the isolated frontal bone.

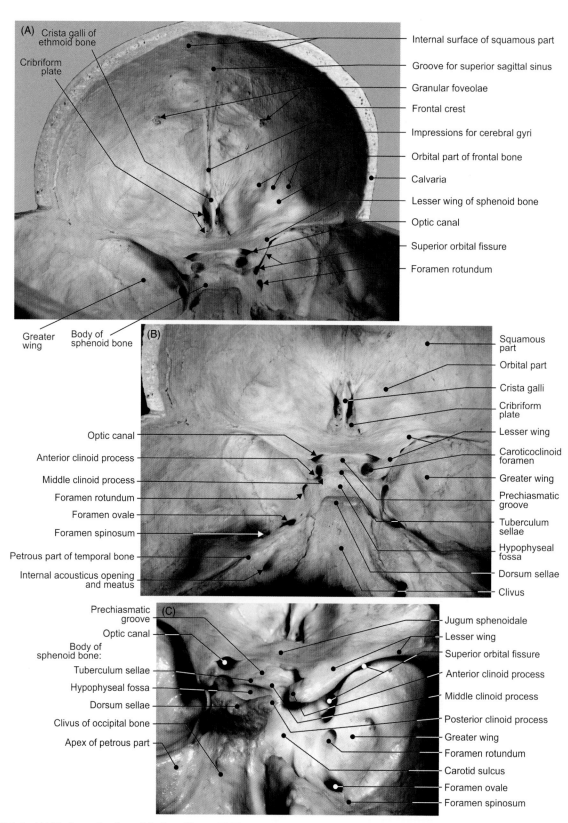

(A) Crista galli of ethmoid bone

Cribriform plate

Internal surface of squamous part

Groove for superior sagittal sinus

Granular foveolae

Frontal crest

Impressions for cerebral gyri

Orbital part of frontal bone

Calvaria

Lesser wing of sphenoid bone

Optic canal

Superior orbital fissure

Foramen rotundum

Greater wing

Body of sphenoid bone

(B)

Squamous part

Orbital part

Crista galli

Cribriform plate

Lesser wing

Caroticoclinoid foramen

Greater wing

Prechiasmatic groove

Tuberculum sellae

Hypophyseal fossa

Dorsum sellae

Clivus

Optic canal

Anterior clinoid process

Middle clinoid process

Foramen rotundum

Foramen ovale

Foramen spinosum

Petrous part of temporal bone

Internal acousticus opening and meatus

(C)

Prechiasmatic groove

Optic canal

Body of sphenoid bone:

Tuberculum sellae

Hypophyseal fossa

Dorsum sellae

Clivus of occipital bone

Apex of petrous part

Jugum sphenoidale

Lesser wing

Superior orbital fissure

Anterior clinoid process

Middle clinoid process

Posterior clinoid process

Greater wing

Foramen rotundum

Carotid sulcus

Foramen ovale

Foramen spinosum

FIGURE 6.4   (A) The internal surface of the frontal bone. (B) The anterior cranial fossa and (C) middle cranial fossa.

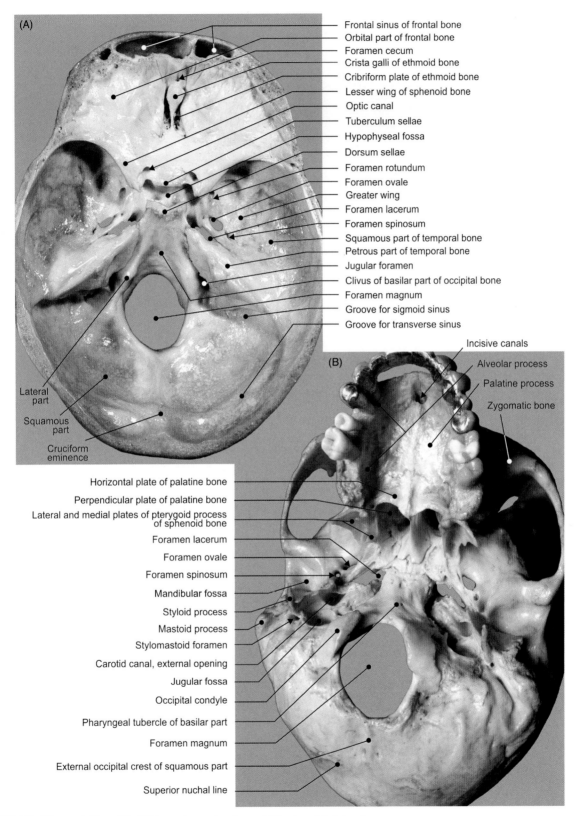

(A)

Frontal sinus of frontal bone
Orbital part of frontal bone
Foramen cecum
Crista galli of ethmoid bone
Cribriform plate of ethmoid bone
Lesser wing of sphenoid bone
Optic canal
Tuberculum sellae
Hypophyseal fossa
Dorsum sellae
Foramen rotundum
Foramen ovale
Greater wing
Foramen lacerum
Foramen spinosum
Squamous part of temporal bone
Petrous part of temporal bone
Jugular foramen
Clivus of basilar part of occipital bone
Foramen magnum
Groove for sigmoid sinus
Groove for transverse sinus

Lateral part
Squamous part
Cruciform eminence

Incisive canals
Alveolar process
Palatine process
Zygomatic bone

(B)

Horizontal plate of palatine bone
Perpendicular plate of palatine bone
Lateral and medial plates of pterygoid process of sphenoid bone
Foramen lacerum
Foramen ovale
Foramen spinosum
Mandibular fossa
Styloid process
Mastoid process
Stylomastoid foramen
Carotid canal, external opening
Jugular fossa
Occipital condyle
Pharyngeal tubercle of basilar part
Foramen magnum
External occipital crest of squamous part
Superior nuchal line

FIGURE 6.5   The cranial base. The (A) internal, superior view and (B) external, inferior view.

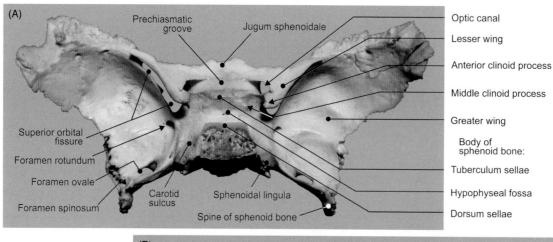

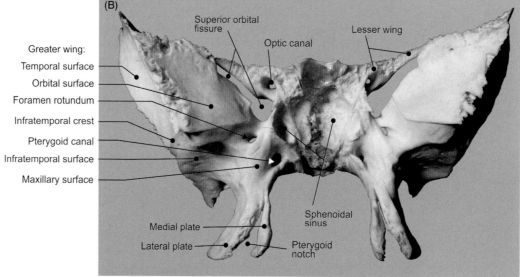

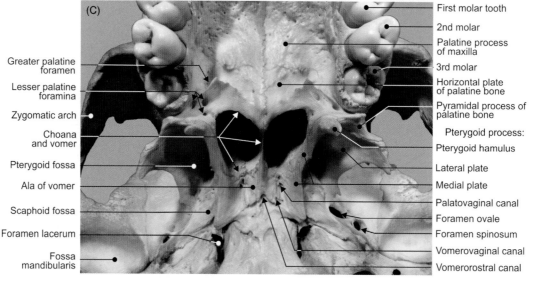

FIGURE 6.6 The isolated sphenoid bone: (A) view from above and (B) view from the front. (C) External, inferior aspect of the cranial base.

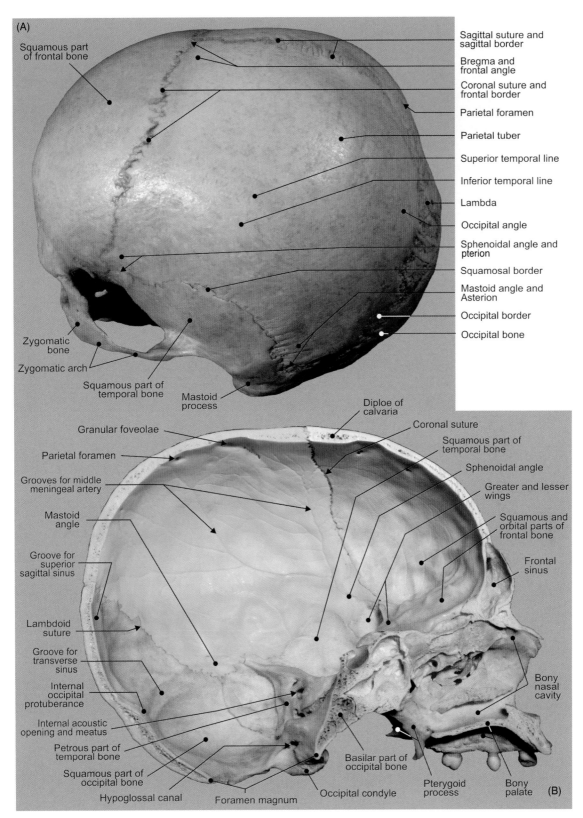

FIGURE 6.7 The (A) external and (B) internal aspects of the parietal bone.

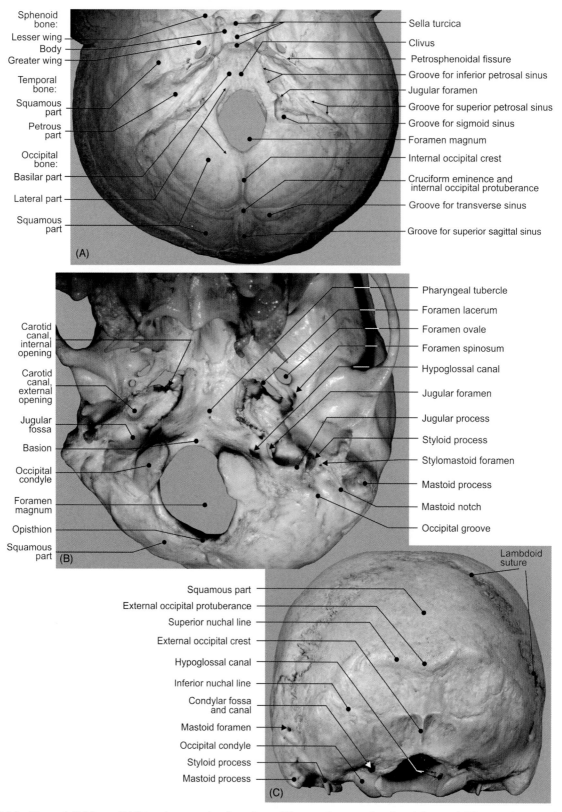

FIGURE 6.8 **The occipital bone.** (A) Internal aspect, view from above; (B) external aspect, view from below; and (C) external aspect, view from behind.

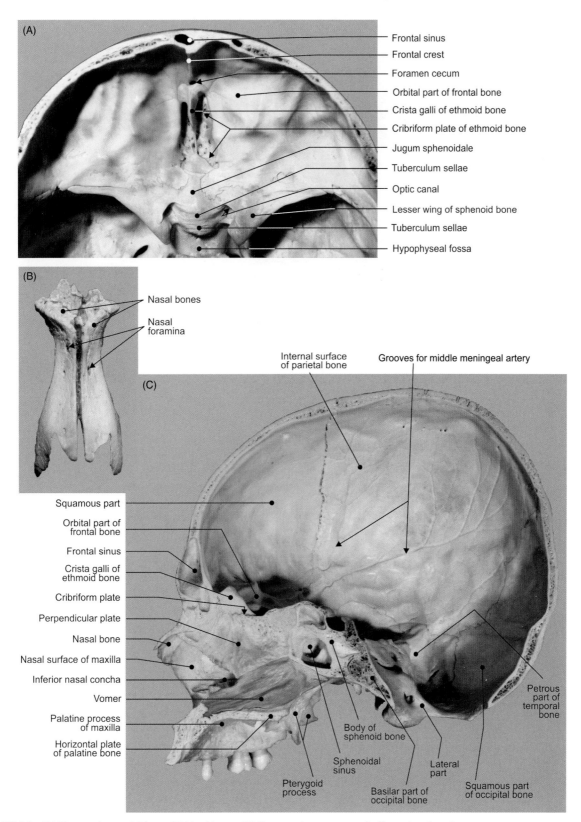

FIGURE 6.9   (A) The anterior cranial fossa. (B) Nasal bones. (C) Bony nasal septum on sagittally sectioned cranium.

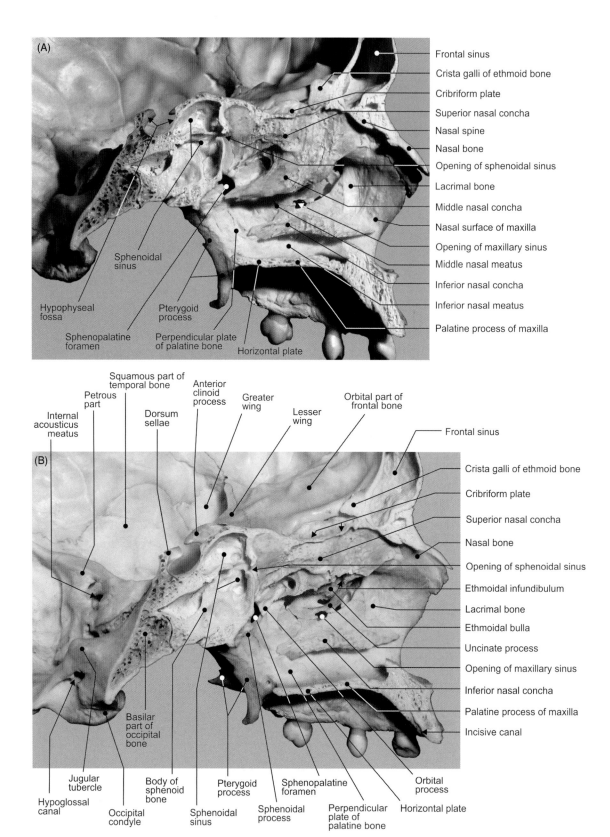

(A)

Frontal sinus

Crista galli of ethmoid bone

Cribriform plate

Superior nasal concha

Nasal spine

Nasal bone

Opening of sphenoidal sinus

Lacrimal bone

Middle nasal concha

Nasal surface of maxilla

Opening of maxillary sinus

Middle nasal meatus

Inferior nasal concha

Inferior nasal meatus

Palatine process of maxilla

Sphenoidal sinus

Hypophyseal fossa

Pterygoid process

Sphenopalatine foramen

Perpendicular plate of palatine bone

Horizontal plate

(B)

Squamous part of temporal bone

Petrous part

Internal acousticus meatus

Dorsum sellae

Anterior clinoid process

Greater wing

Lesser wing

Orbital part of frontal bone

Frontal sinus

Crista galli of ethmoid bone

Cribriform plate

Superior nasal concha

Nasal bone

Opening of sphenoidal sinus

Ethmoidal infundibulum

Lacrimal bone

Ethmoidal bulla

Uncinate process

Opening of maxillary sinus

Inferior nasal concha

Palatine process of maxilla

Incisive canal

Jugular tubercle

Hypoglossal canal

Occipital condyle

Basilar part of occipital bone

Body of sphenoid bone

Sphenoidal sinus

Pterygoid process

Sphenopalatine foramen

Sphenoidal process

Perpendicular plate of palatine bone

Orbital process

Horizontal plate

FIGURE 6.10   (A–B) Different views displaying the midsagittal sections of the bony nasal cavity.

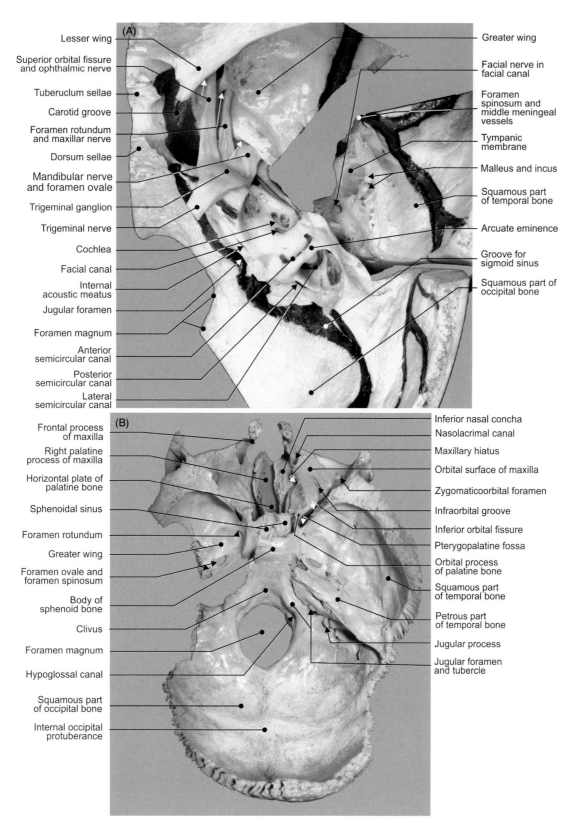

FIGURE 6.11 (A) View from the above of section and dissection of the temporal bone. (B) The internal aspect of the cranial base. Following bones were removed: frontal, ethmoid, part of sphenoid, left temporal, parietal, and small facial.

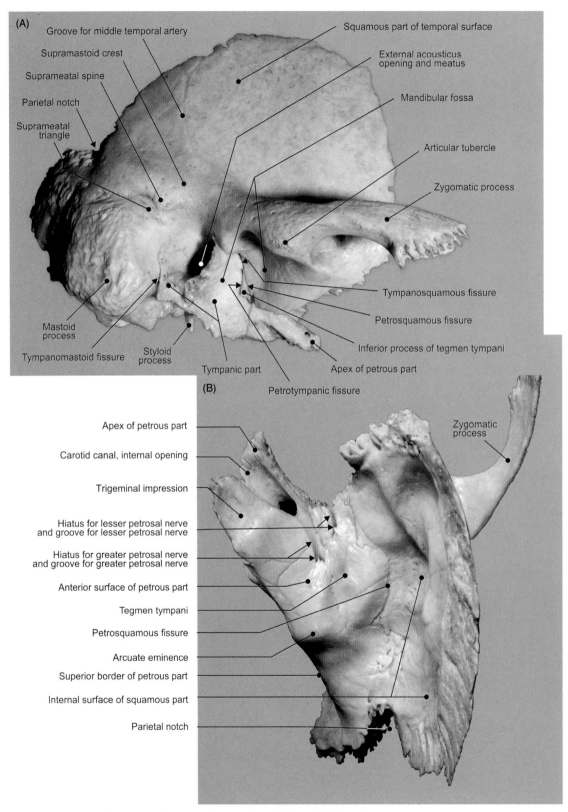

(A)
Groove for middle temporal artery
Supramastoid crest
Suprameatal spine
Parietal notch
Suprameatal triangle
Mastoid process
Tympanomastoid fissure
Styloid process
Tympanic part
Petrotympanic fissure
Squamous part of temporal surface
External acousticus opening and meatus
Mandibular fossa
Articular tubercle
Zygomatic process
Tympanosquamous fissure
Petrosquamous fissure
Inferior process of tegmen tympani
Apex of petrous part

(B)
Apex of petrous part
Carotid canal, internal opening
Trigeminal impression
Hiatus for lesser petrosal nerve and groove for lesser petrosal nerve
Hiatus for greater petrosal nerve and groove for greater petrosal nerve
Anterior surface of petrous part
Tegmen tympani
Petrosquamous fissure
Arcuate eminence
Superior border of petrous part
Internal surface of squamous part
Parietal notch
Zygomatic process

FIGURE 6.12   The isolated right temporal bone. (A) View from lateral side and (B) from above.

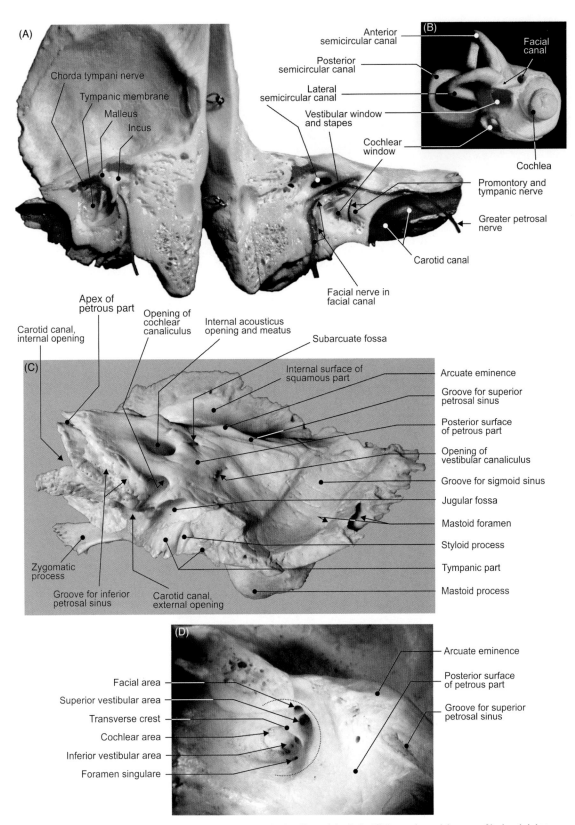

(A) Chorda tympani nerve
Tympanic membrane
Malleus
Incus

(B) Anterior semicircular canal
Posterior semicircular canal
Lateral semicircular canal
Vestibular window and stapes
Cochlear window
Facial canal
Cochlea
Promontory and tympanic nerve
Greater petrosal nerve
Carotid canal
Facial nerve in facial canal

(C) Apex of petrous part
Carotid canal, internal opening
Opening of cochlear canaliculus
Internal acousticus opening and meatus
Subarcuate fossa
Internal surface of squamous part
Arcuate eminence
Groove for superior petrosal sinus
Posterior surface of petrous part
Opening of vestibular canaliculus
Groove for sigmoid sinus
Jugular fossa
Mastoid foramen
Styloid process
Tympanic part
Mastoid process
Zygomatic process
Groove for inferior petrosal sinus
Carotid canal, external opening

(D) Facial area
Superior vestibular area
Transverse crest
Cochlear area
Inferior vestibular area
Foramen singulare
Arcuate eminence
Posterior surface of petrous part
Groove for superior petrosal sinus

FIGURE 6.13 (A) The longitudinal section of temporal bone and (B) isolated bony labyrinth. (C) Internal, cranial aspect of isolated right temporal bone. (D) Dissection of fundus of the right internal acoustic meatus.

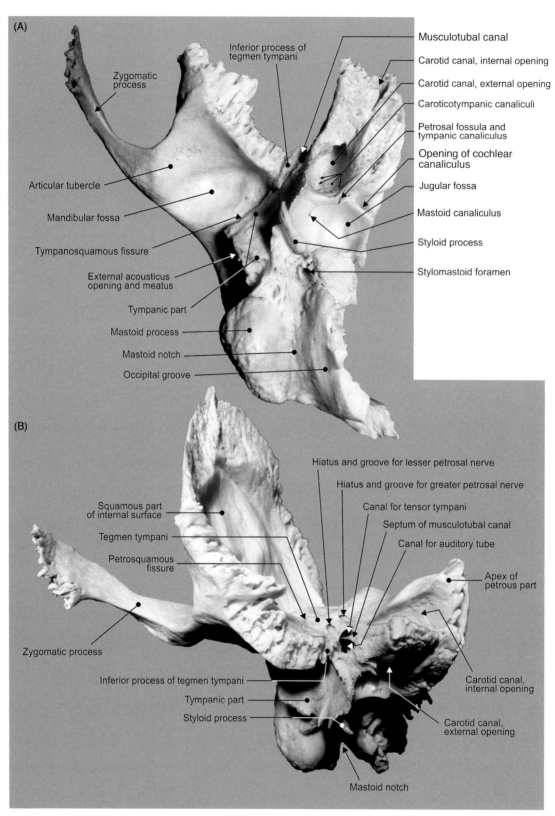

FIGURE 6.14 **The right temporal bone.** (A) View from below and (B) from the front.

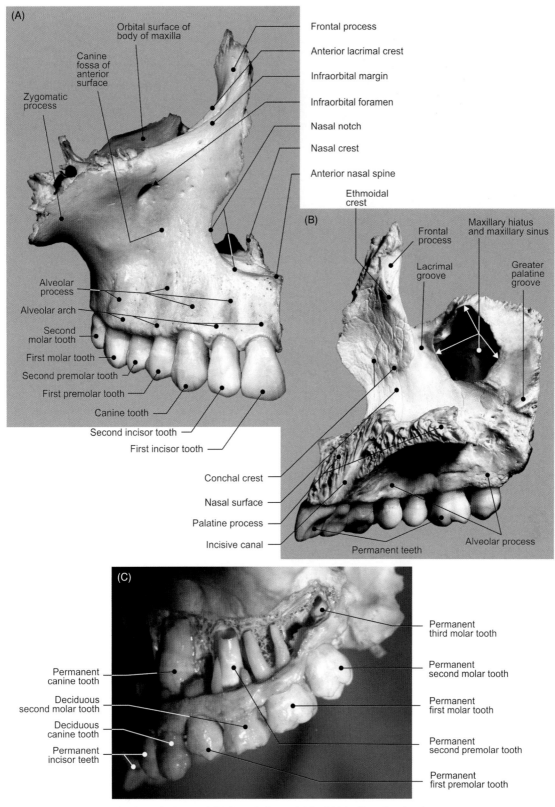

FIGURE 6.15   The maxilla. (A) View from the front and (B) from the internal side. (C) Maxillary dental arch.

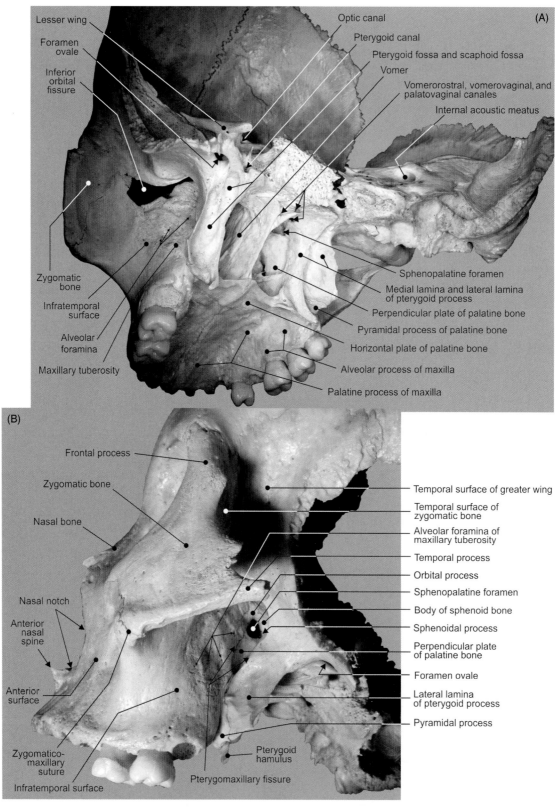

FIGURE 6.16 **The palatine bone.** (A) View from behind. (B) Pterygopalatine, or sphenopalatine fossa.

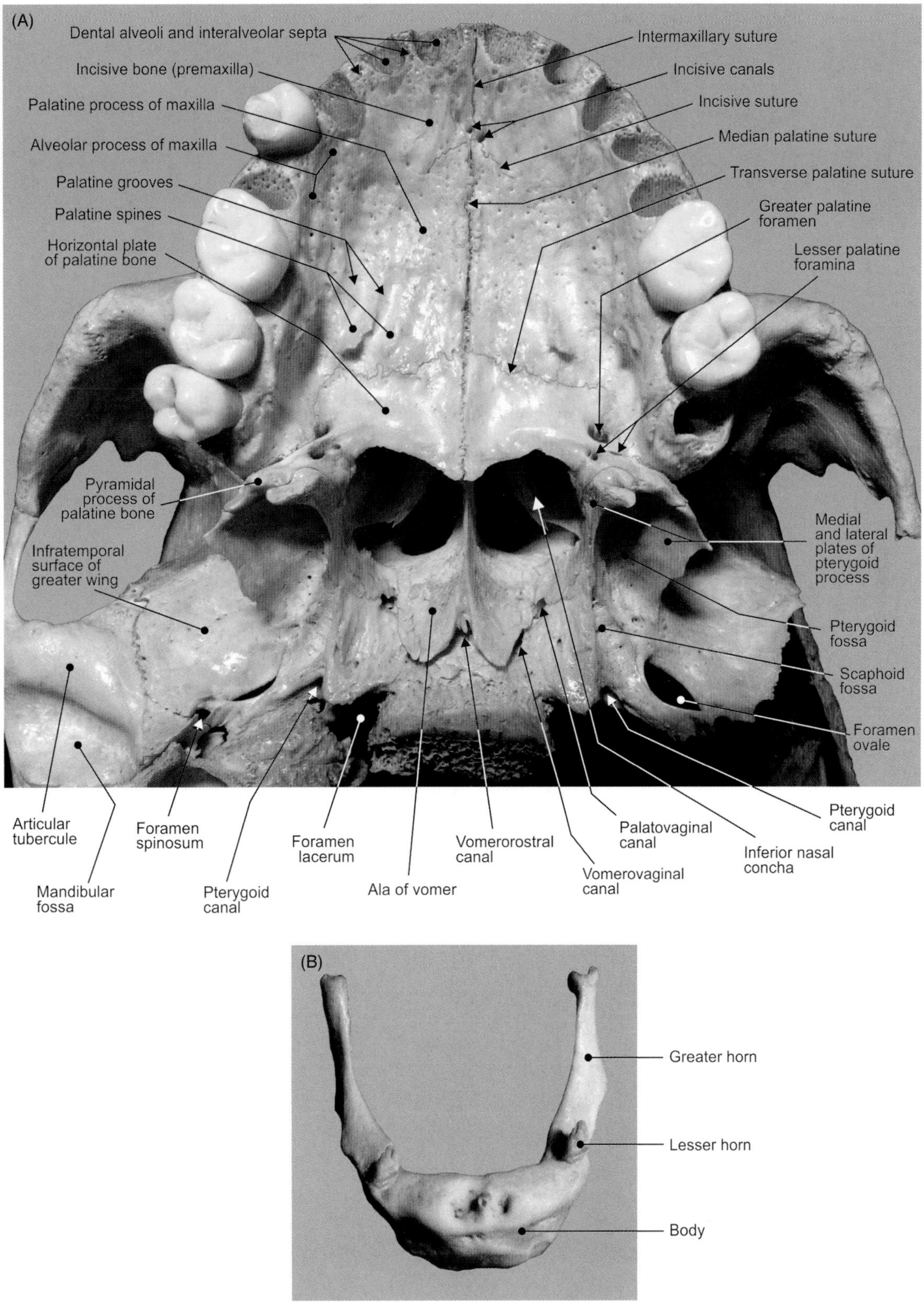

(A)

Dental alveoli and interalveolar septa

Incisive bone (premaxilla)

Palatine process of maxilla

Alveolar process of maxilla

Palatine grooves

Palatine spines

Horizontal plate of palatine bone

Intermaxillary suture

Incisive canals

Incisive suture

Median palatine suture

Transverse palatine suture

Greater palatine foramen

Lesser palatine foramina

Pyramidal process of palatine bone

Infratemporal surface of greater wing

Medial and lateral plates of pterygoid process

Pterygoid fossa

Scaphoid fossa

Foramen ovale

Articular tubercule

Foramen spinosum

Foramen lacerum

Vomerorostral canal

Palatovaginal canal

Pterygoid canal

Inferior nasal concha

Mandibular fossa

Pterygoid canal

Ala of vomer

Vomerovaginal canal

(B)

Greater horn

Lesser horn

Body

FIGURE 6.17 (A) Inferior view of the bony, hard palate. (B) Superior view of the hyoid bone.

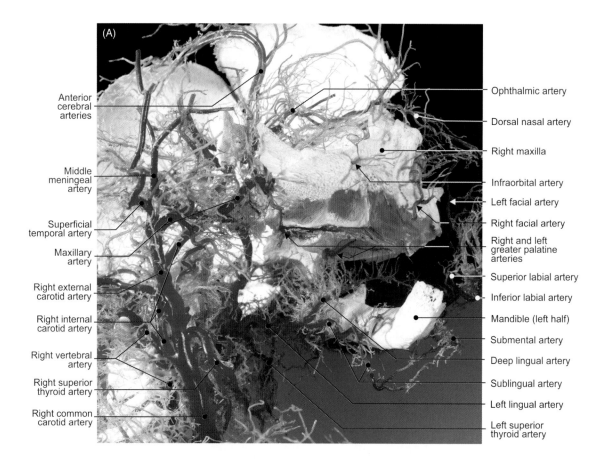

Anterior cerebral arteries

Middle meningeal artery

Superficial temporal artery

Maxillary artery

Right external carotid artery

Right internal carotid artery

Right vertebral artery

Right superior thyroid artery

Right common carotid artery

Ophthalmic artery

Dorsal nasal artery

Right maxilla

Infraorbital artery

Left facial artery

Right facial artery

Right and left greater palatine arteries

Superior labial artery

Inferior labial artery

Mandible (left half)

Submental artery

Deep lingual artery

Sublingual artery

Left lingual artery

Left superior thyroid artery

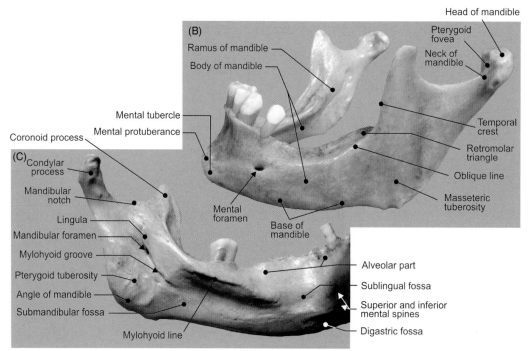

Head of mandible

Pterygoid fovea

Neck of mandible

Ramus of mandible

Body of mandible

Mental tubercle

Mental protuberance

Coronoid process

Condylar process

Mandibular notch

Lingula

Mandibular foramen

Mylohyoid groove

Pterygoid tuberosity

Angle of mandible

Submandibular fossa

Mylohyoid line

Mental foramen

Base of mandible

Temporal crest

Retromolar triangle

Oblique line

Masseteric tuberosity

Alveolar part

Sublingual fossa

Superior and inferior mental spines

Digastric fossa

FIGURE 6.18 (A) The fetal arteries of the head and neck: inferior view from the right side. (B–C) Mandible: view of the external and internal side.

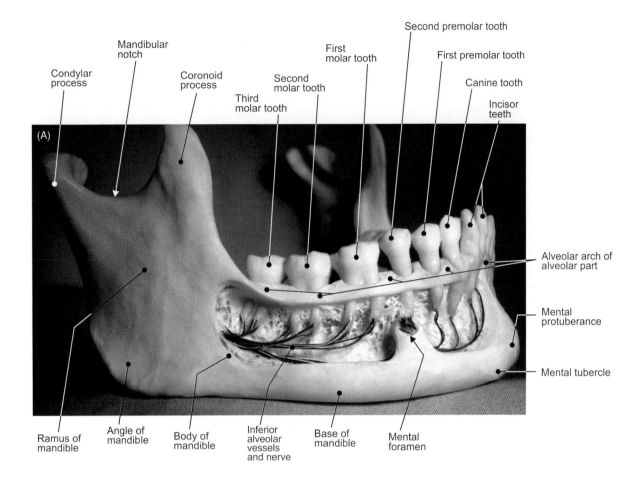

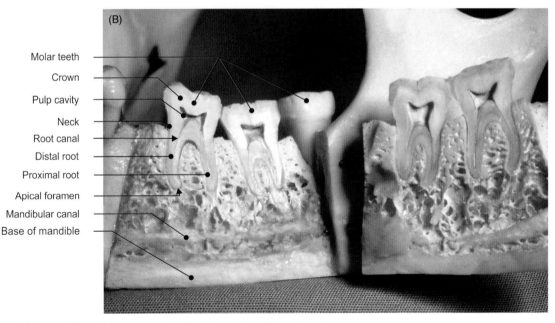

FIGURE 6.19   The mandible. (A) Dental arch and (B) sagittal section of the molar region.

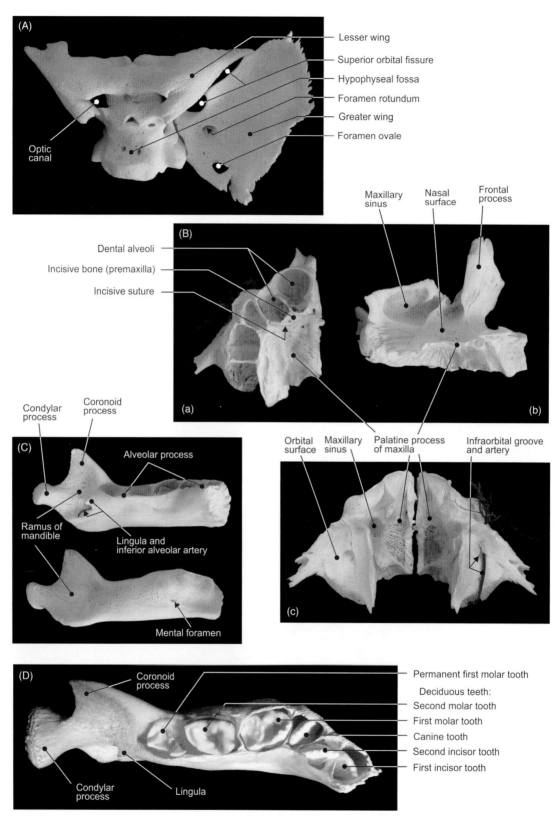

FIGURE 6.20 (A) Fetal sphenoid bone: superior view. (B) Fetal maxilla: (a) inferior, (b) internal, and (c) superior views. (C–D) Different views of the fetal mandible.

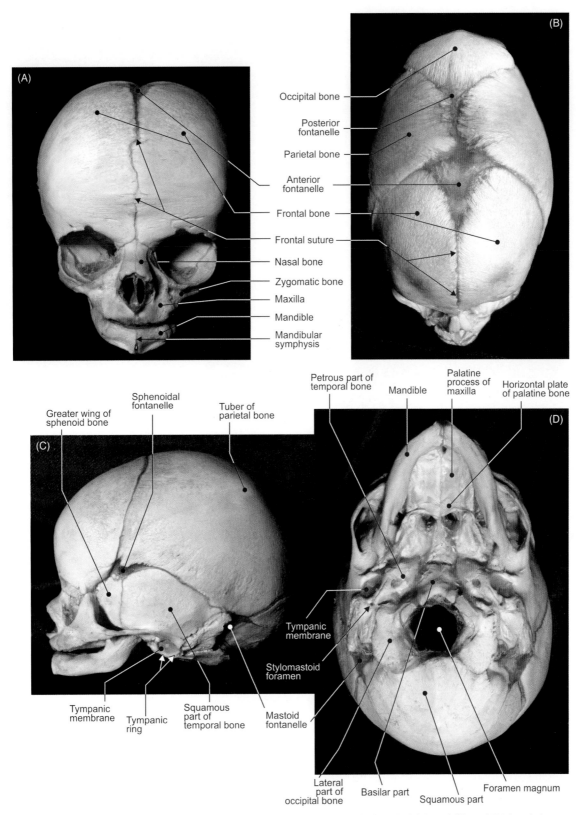

FIGURE 6.21 **The fetal skull with mandible.** (A) View from the front, (B) from above, (C) from the left lateral side, and (D) from below.

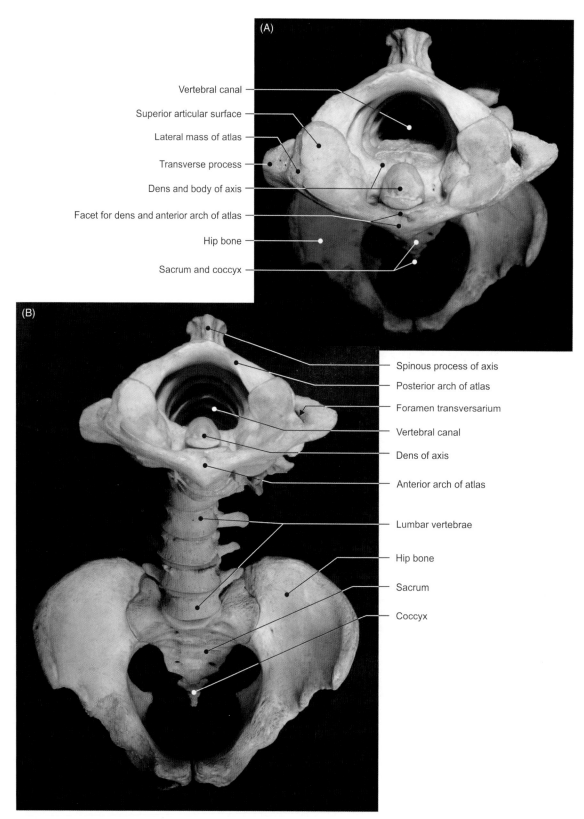

FIGURE 6.22 The vertebral column. (A) View from above and (B) from the front.

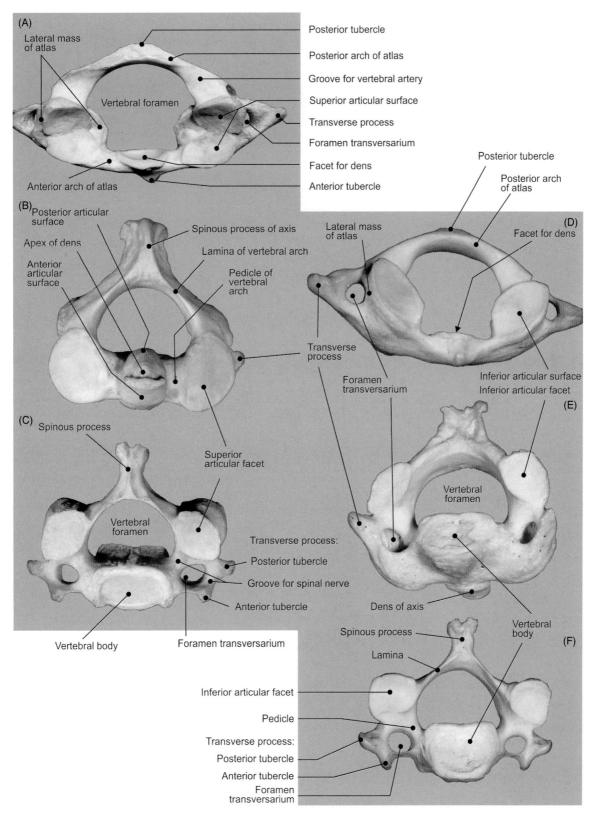

FIGURE 6.23   Atlas, axis, and C3 vertebra. (A–C) Views from above and (D–F) from below.

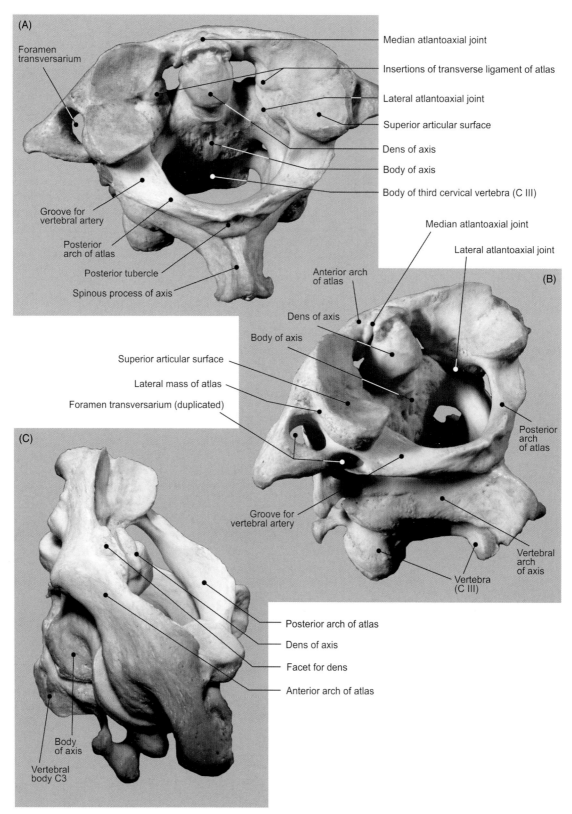

FIGURE 6.24   Atlas, axis, and vertebra C3. (A) View from behind and (B–C) from the left lateral side.

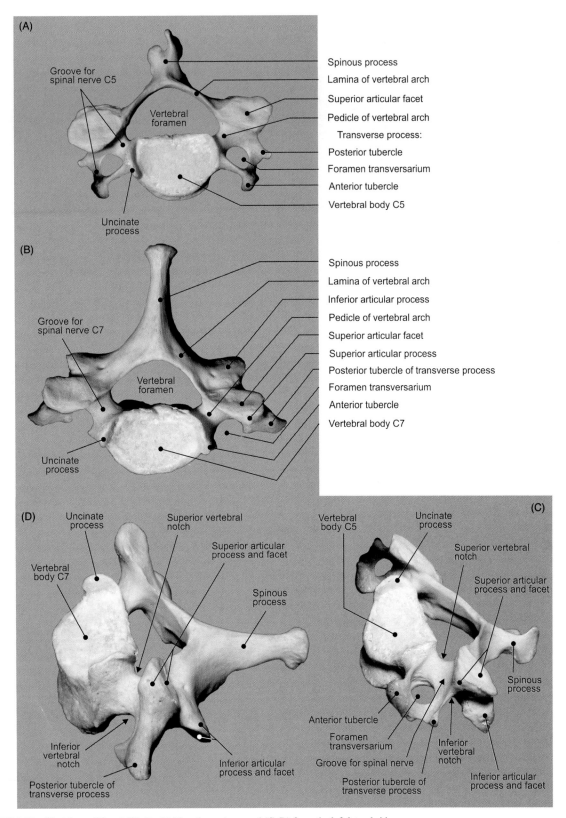

FIGURE 6.25 **Vertebrae C5 and C7.** (A–B) View from above and (C–D) from the left lateral side.

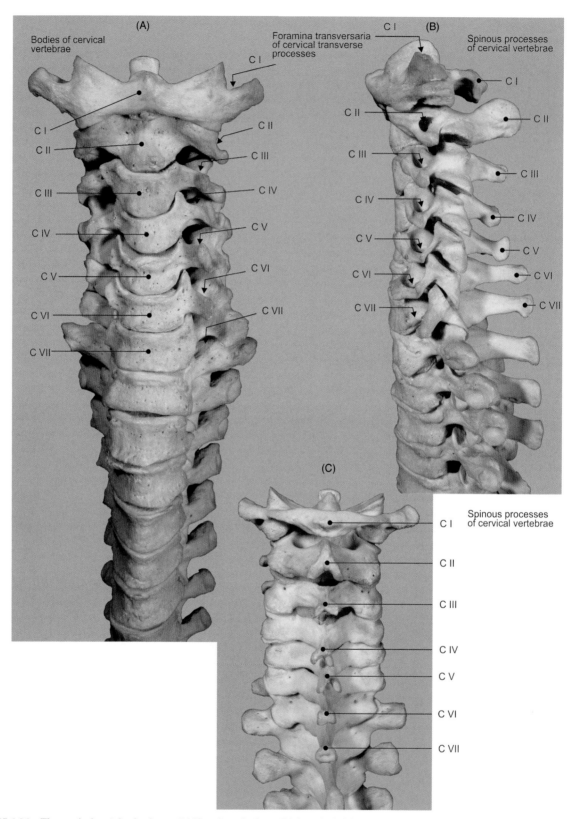

FIGURE 6.26 The cervical vertebral column. (A) View from the front, (B) from the left lateral side, and (C) from behind.

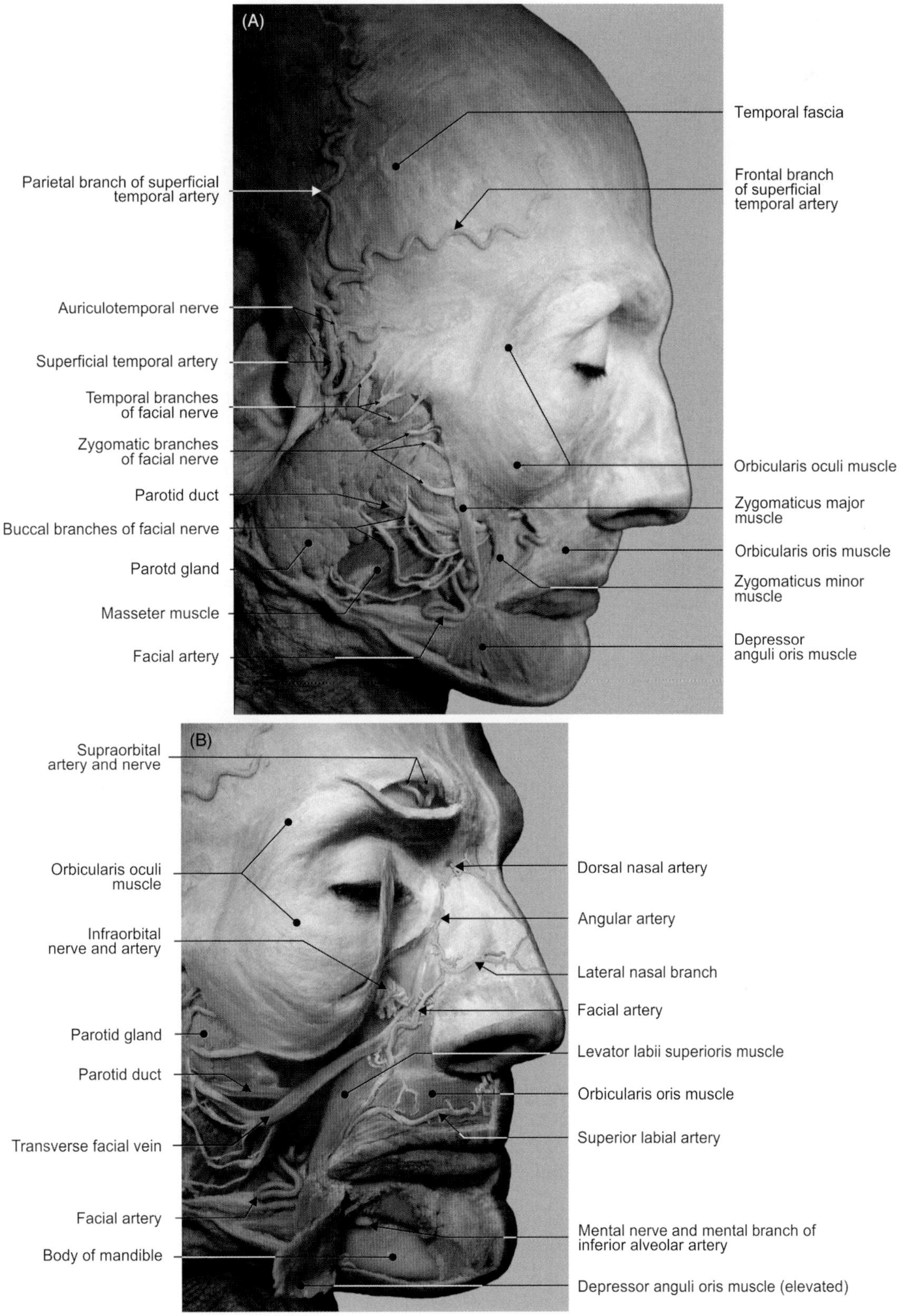

FIGURE 6.27 The facial and parotid regions. (A) Superficial muscular layer and (A–B) nerves and blood vessels.

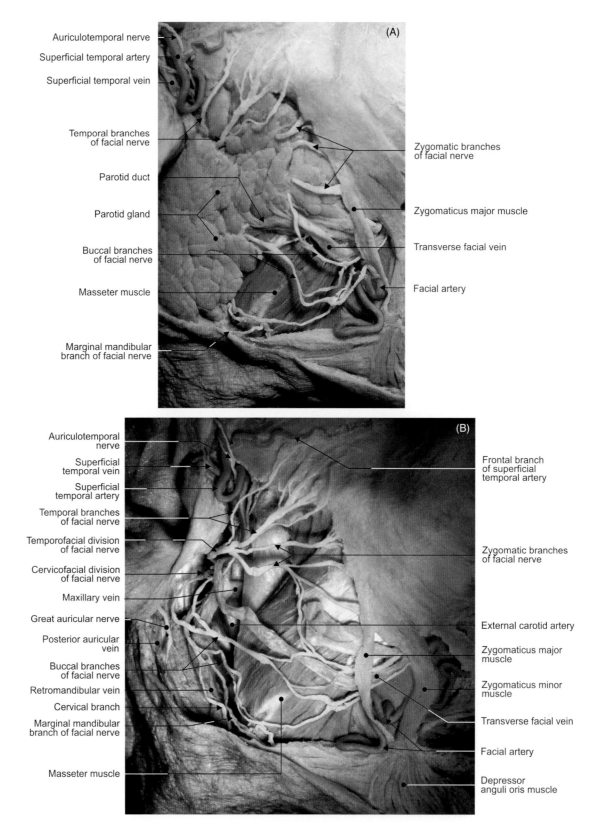

Auriculotemporal nerve
Superficial temporal artery
Superficial temporal vein

Temporal branches
of facial nerve

Parotid duct

Parotid gland

Buccal branches
of facial nerve

Masseter muscle

Marginal mandibular
branch of facial nerve

(A)

Zygomatic branches
of facial nerve

Zygomaticus major muscle

Transverse facial vein

Facial artery

Auriculotemporal
nerve
Superficial
temporal vein
Superficial
temporal artery
Temporal branches
of facial nerve
Temporofacial division
of facial nerve
Cervicofacial division
of facial nerve
Maxillary vein
Great auricular nerve
Posterior auricular
vein
Buccal branches
of facial nerve
Retromandibular vein
Cervical branch
Marginal mandibular
branch of facial nerve

Masseter muscle

(B)

Frontal branch
of superficial
temporal artery

Zygomatic branches
of facial nerve

External carotid artery

Zygomaticus major
muscle

Zygomaticus minor
muscle

Transverse facial vein

Facial artery

Depressor
anguli oris muscle

FIGURE 6.28 The parotidomasseteric region. (A) With parotid gland and (B) after the gland has been removed.

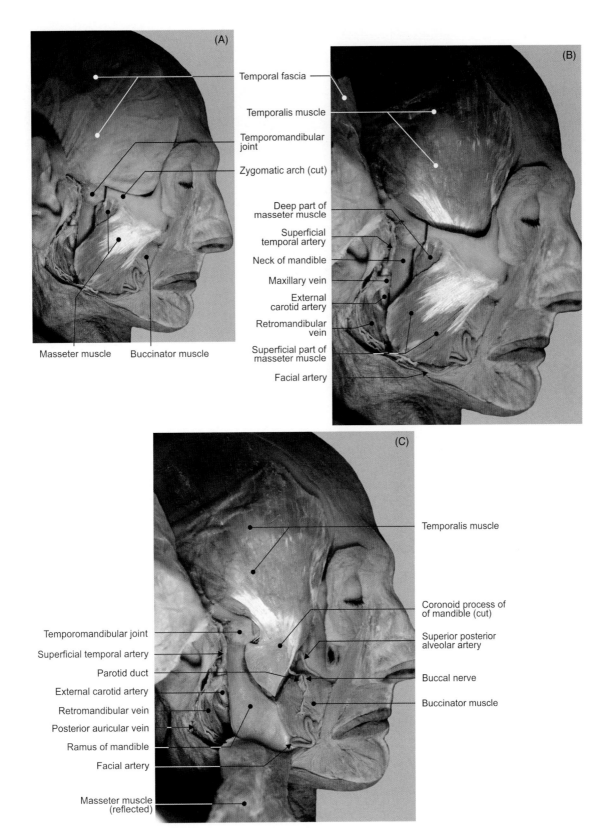

(A)

Temporal fascia

Temporalis muscle

Temporomandibular joint

Zygomatic arch (cut)

Deep part of masseter muscle

Superficial temporal artery

Neck of mandible

Maxillary vein

External carotid artery

Retromandibular vein

Superficial part of masseter muscle

Facial artery

Masseter muscle   Buccinator muscle

(B)

(C)

Temporalis muscle

Coronoid process of of mandible (cut)

Superior posterior alveolar artery

Buccal nerve

Buccinator muscle

Temporomandibular joint

Superficial temporal artery

Parotid duct

External carotid artery

Retromandibular vein

Posterior auricular vein

Ramus of mandible

Facial artery

Masseter muscle (reflected)

FIGURE 6.29   **The muscles of mastication.** (A) Masseter and (B–C) temporalis.

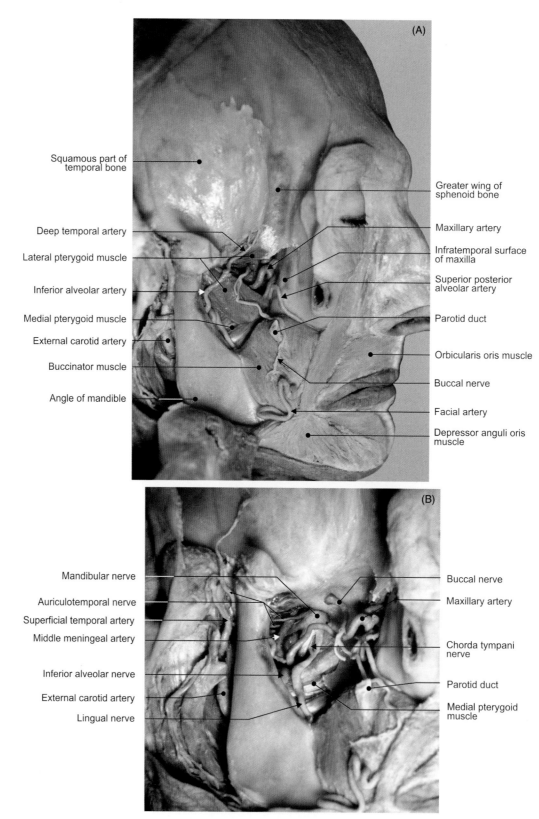

Squamous part of temporal bone

Greater wing of sphenoid bone

Deep temporal artery

Maxillary artery

Lateral pterygoid muscle

Infratemporal surface of maxilla

Inferior alveolar artery

Superior posterior alveolar artery

Medial pterygoid muscle

Parotid duct

External carotid artery

Orbicularis oris muscle

Buccinator muscle

Buccal nerve

Angle of mandible

Facial artery

Depressor anguli oris muscle

Mandibular nerve

Buccal nerve

Auriculotemporal nerve

Maxillary artery

Superficial temporal artery

Middle meningeal artery

Chorda tympani nerve

Inferior alveolar nerve

Parotid duct

External carotid artery

Medial pterygoid muscle

Lingual nerve

FIGURE 6.30 The infratemporal fossa. (A) Structures in superficial and (B) deep layers.

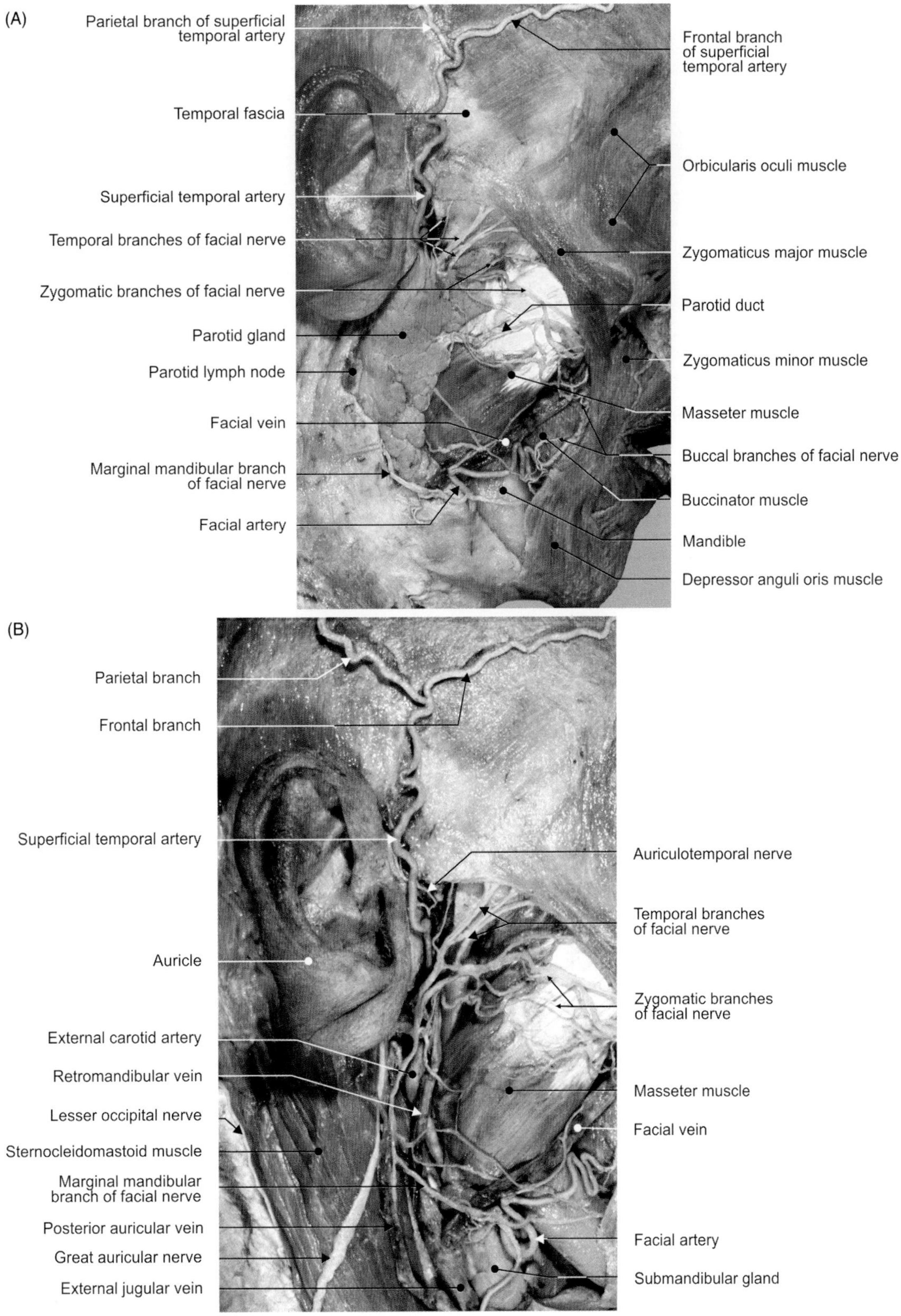

(A)

Parietal branch of superficial temporal artery

Temporal fascia

Superficial temporal artery

Temporal branches of facial nerve

Zygomatic branches of facial nerve

Parotid gland

Parotid lymph node

Facial vein

Marginal mandibular branch of facial nerve

Facial artery

Frontal branch of superficial temporal artery

Orbicularis oculi muscle

Zygomaticus major muscle

Parotid duct

Zygomaticus minor muscle

Masseter muscle

Buccal branches of facial nerve

Buccinator muscle

Mandible

Depressor anguli oris muscle

(B)

Parietal branch

Frontal branch

Superficial temporal artery

Auricle

External carotid artery

Retromandibular vein

Lesser occipital nerve

Sternocleidomastoid muscle

Marginal mandibular branch of facial nerve

Posterior auricular vein

Great auricular nerve

External jugular vein

Auriculotemporal nerve

Temporal branches of facial nerve

Zygomatic branches of facial nerve

Masseter muscle

Facial vein

Facial artery

Submandibular gland

FIGURE 6.31 **The parotidomasseteric region.** (A) With parotid gland and (B) after the gland has been removed.

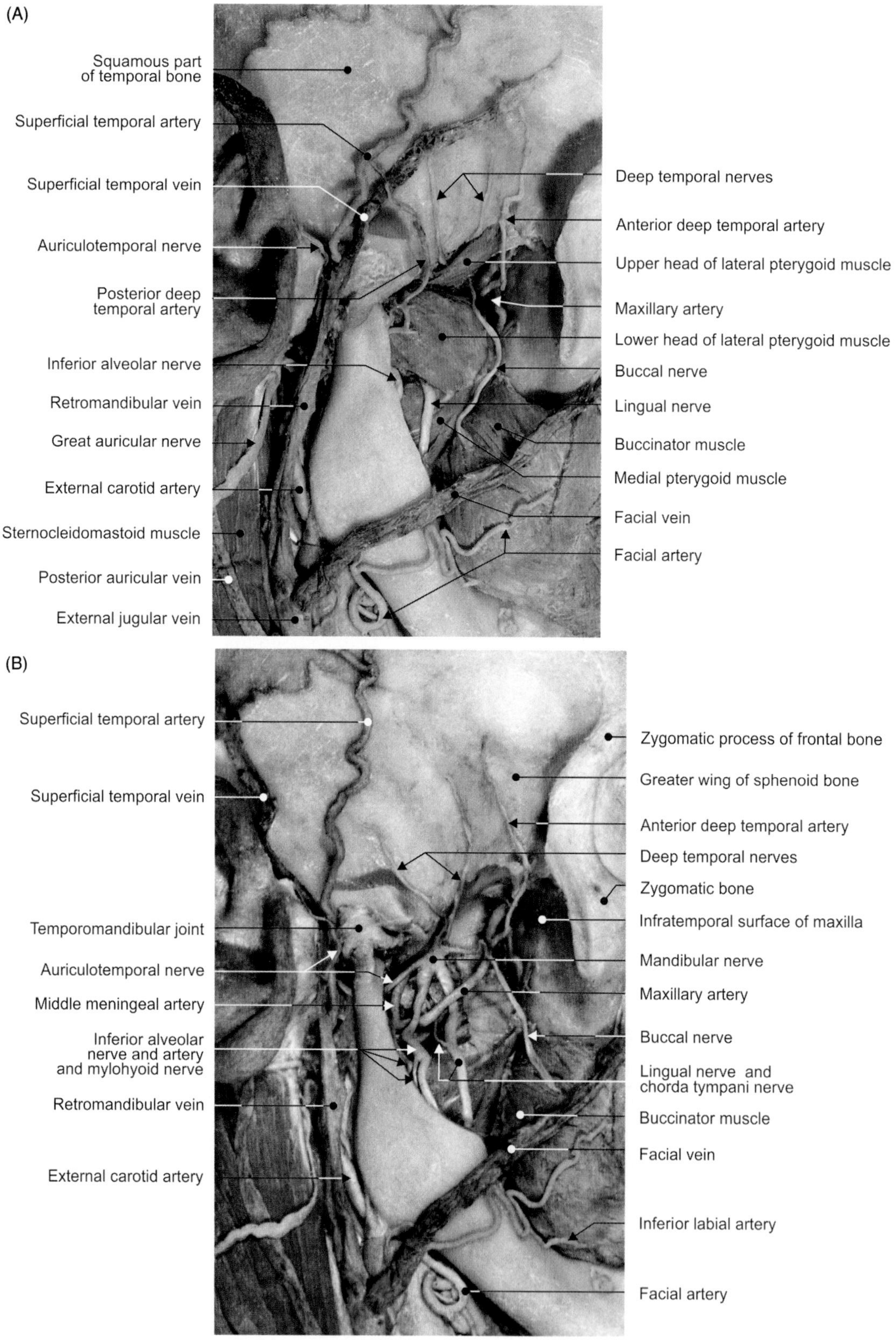

(A)

Squamous part
of temporal bone

Superficial temporal artery

Superficial temporal vein

Auriculotemporal nerve

Posterior deep
temporal artery

Inferior alveolar nerve

Retromandibular vein

Great auricular nerve

External carotid artery

Sternocleidomastoid muscle

Posterior auricular vein

External jugular vein

Deep temporal nerves

Anterior deep temporal artery

Upper head of lateral pterygoid muscle

Maxillary artery

Lower head of lateral pterygoid muscle

Buccal nerve

Lingual nerve

Buccinator muscle

Medial pterygoid muscle

Facial vein

Facial artery

(B)

Superficial temporal artery

Superficial temporal vein

Temporomandibular joint

Auriculotemporal nerve

Middle meningeal artery

Inferior alveolar
nerve and artery
and mylohyoid nerve

Retromandibular vein

External carotid artery

Zygomatic process of frontal bone

Greater wing of sphenoid bone

Anterior deep temporal artery

Deep temporal nerves

Zygomatic bone

Infratemporal surface of maxilla

Mandibular nerve

Maxillary artery

Buccal nerve

Lingual nerve and
chorda tympani nerve

Buccinator muscle

Facial vein

Inferior labial artery

Facial artery

FIGURE 6.32   The infratemporal fossa. Structures in the (A) superficial and (B) deep layers.

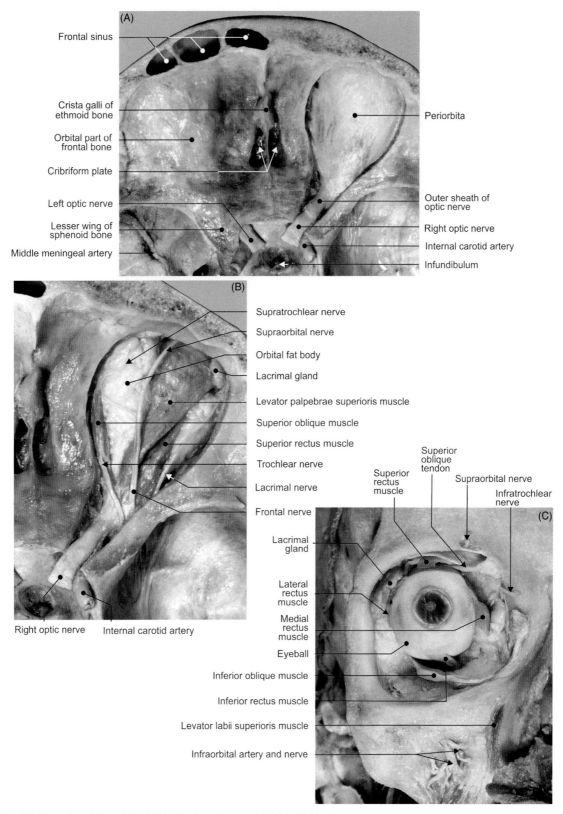

(A)

Frontal sinus

Crista galli of ethmoid bone

Orbital part of frontal bone

Cribriform plate

Left optic nerve

Lesser wing of sphenoid bone

Middle meningeal artery

Periorbita

Outer sheath of optic nerve

Right optic nerve

Internal carotid artery

Infundibulum

(B)

Supratrochlear nerve

Supraorbital nerve

Orbital fat body

Lacrimal gland

Levator palpebrae superioris muscle

Superior oblique muscle

Superior rectus muscle

Trochlear nerve

Lacrimal nerve

Frontal nerve

Right optic nerve   Internal carotid artery

Lacrimal gland

Lateral rectus muscle

Medial rectus muscle

Eyeball

Inferior oblique muscle

Inferior rectus muscle

Levator labii superioris muscle

Infraorbital artery and nerve

Superior rectus muscle

Superior oblique tendon

Supraorbital nerve

Infratrochlear nerve

(C)

FIGURE 6.33   Dissection of the orbit. (A–B) View from above and (C) from the front.

(A)

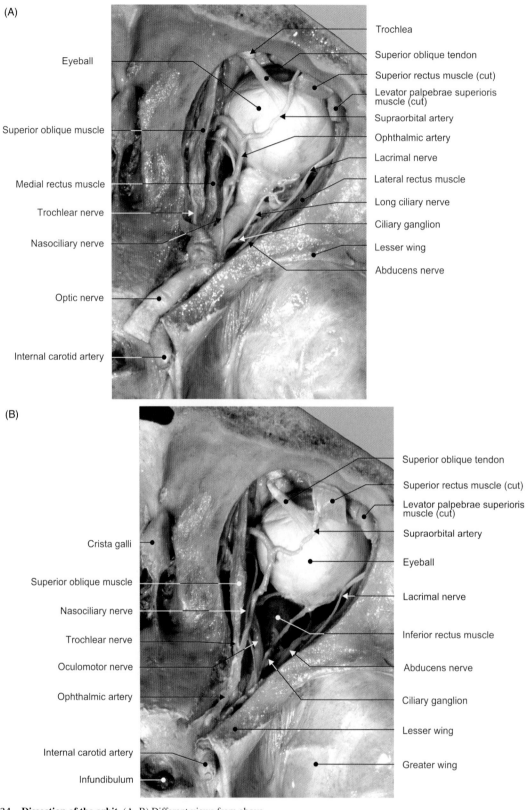

Eyeball

Trochlea

Superior oblique tendon

Superior rectus muscle (cut)

Levator palpebrae superioris muscle (cut)

Superior oblique muscle

Supraorbital artery

Ophthalmic artery

Lacrimal nerve

Medial rectus muscle

Lateral rectus muscle

Trochlear nerve

Long ciliary nerve

Nasociliary nerve

Ciliary ganglion

Lesser wing

Abducens nerve

Optic nerve

Internal carotid artery

(B)

Crista galli

Superior oblique tendon

Superior rectus muscle (cut)

Levator palpebrae superioris muscle (cut)

Superior oblique muscle

Supraorbital artery

Nasociliary nerve

Eyeball

Trochlear nerve

Oculomotor nerve

Lacrimal nerve

Ophthalmic artery

Inferior rectus muscle

Abducens nerve

Internal carotid artery

Ciliary ganglion

Infundibulum

Lesser wing

Greater wing

FIGURE 6.34   Dissection of the orbit. (A–B) Different views from above.

(A)

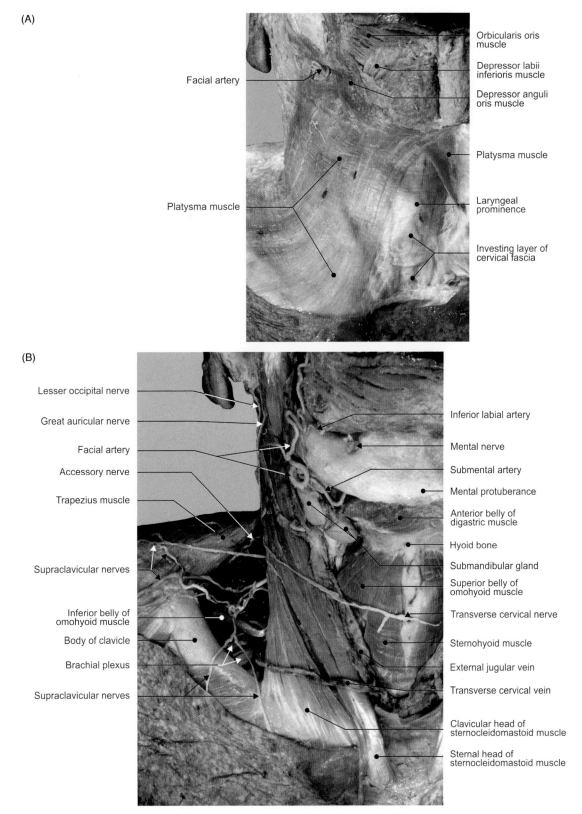

Orbicularis oris muscle

Depressor labii inferioris muscle

Depressor anguli oris muscle

Facial artery

Platysma muscle

Laryngeal prominence

Platysma muscle

Investing layer of cervical fascia

(B)

Lesser occipital nerve

Great auricular nerve

Facial artery

Accessory nerve

Trapezius muscle

Supraclavicular nerves

Inferior belly of omohyoid muscle

Body of clavicle

Brachial plexus

Supraclavicular nerves

Inferior labial artery

Mental nerve

Submental artery

Mental protuberance

Anterior belly of digastric muscle

Hyoid bone

Submandibular gland

Superior belly of omohyoid muscle

Transverse cervical nerve

Sternohyoid muscle

External jugular vein

Transverse cervical vein

Clavicular head of sternocleidomastoid muscle

Sternal head of sternocleidomastoid muscle

FIGURE 6.35 The anterior and lateral cervical regions. (A–B) Different views of the superficial, first layer.

(A)

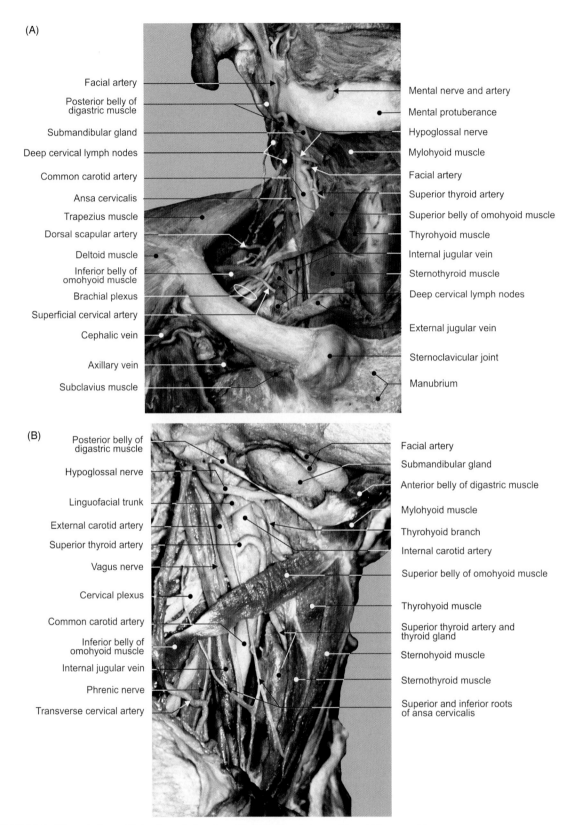

Facial artery

Posterior belly of
digastric muscle

Submandibular gland

Deep cervical lymph nodes

Common carotid artery

Ansa cervicalis

Trapezius muscle

Dorsal scapular artery

Deltoid muscle

Inferior belly of
omohyoid muscle

Brachial plexus

Superficial cervical artery

Cephalic vein

Axillary vein

Subclavius muscle

Mental nerve and artery

Mental protuberance

Hypoglossal nerve

Mylohyoid muscle

Facial artery

Superior thyroid artery

Superior belly of omohyoid muscle

Thyrohyoid muscle

Internal jugular vein

Sternothyroid muscle

Deep cervical lymph nodes

External jugular vein

Sternoclavicular joint

Manubrium

(B)

Posterior belly of
digastric muscle

Hypoglossal nerve

Linguofacial trunk

External carotid artery

Superior thyroid artery

Vagus nerve

Cervical plexus

Common carotid artery

Inferior belly of
omohyoid muscle

Internal jugular vein

Phrenic nerve

Transverse cervical artery

Facial artery

Submandibular gland

Anterior belly of digastric muscle

Mylohyoid muscle

Thyrohyoid branch

Internal carotid artery

Superior belly of omohyoid muscle

Thyrohyoid muscle

Superior thyroid artery and
thyroid gland

Sternohyoid muscle

Sternothyroid muscle

Superior and inferior roots
of ansa cervicalis

FIGURE 6.36 **The anterior and lateral cervical regions.** (A–B) Different views of the middle, second layer.

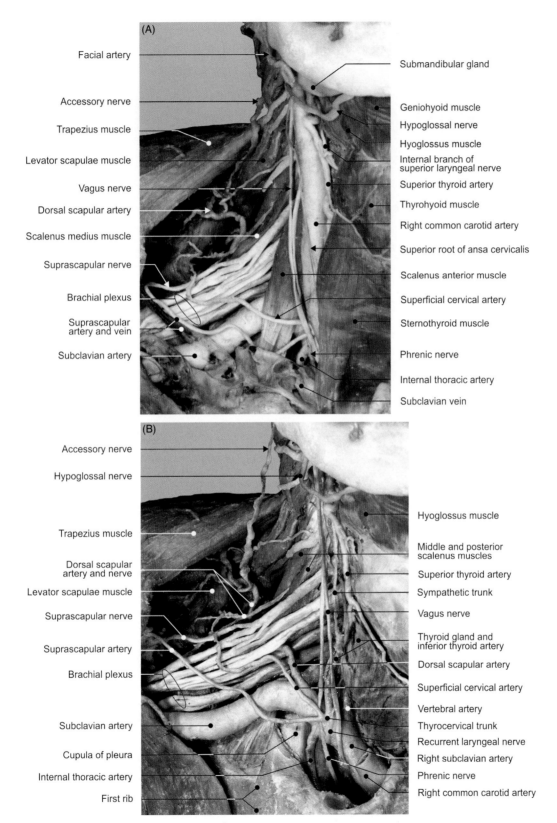

(A)

Facial artery

Accessory nerve

Trapezius muscle

Levator scapulae muscle

Vagus nerve

Dorsal scapular artery

Scalenus medius muscle

Suprascapular nerve

Brachial plexus

Suprascapular artery and vein

Subclavian artery

Submandibular gland

Geniohyoid muscle

Hypoglossal nerve

Hyoglossus muscle

Internal branch of superior laryngeal nerve

Superior thyroid artery

Thyrohyoid muscle

Right common carotid artery

Superior root of ansa cervicalis

Scalenus anterior muscle

Superficial cervical artery

Sternothyroid muscle

Phrenic nerve

Internal thoracic artery

Subclavian vein

(B)

Accessory nerve

Hypoglossal nerve

Trapezius muscle

Dorsal scapular artery and nerve

Levator scapulae muscle

Suprascapular nerve

Suprascapular artery

Brachial plexus

Subclavian artery

Cupula of pleura

Internal thoracic artery

First rib

Hyoglossus muscle

Middle and posterior scalenus muscles

Superior thyroid artery

Sympathetic trunk

Vagus nerve

Thyroid gland and inferior thyroid artery

Dorsal scapular artery

Superficial cervical artery

Vertebral artery

Thyrocervical trunk

Recurrent laryngeal nerve

Right subclavian artery

Phrenic nerve

Right common carotid artery

FIGURE 6.37   The anterior and lateral cervical regions. (A–B) Different views of the deep, third layer.

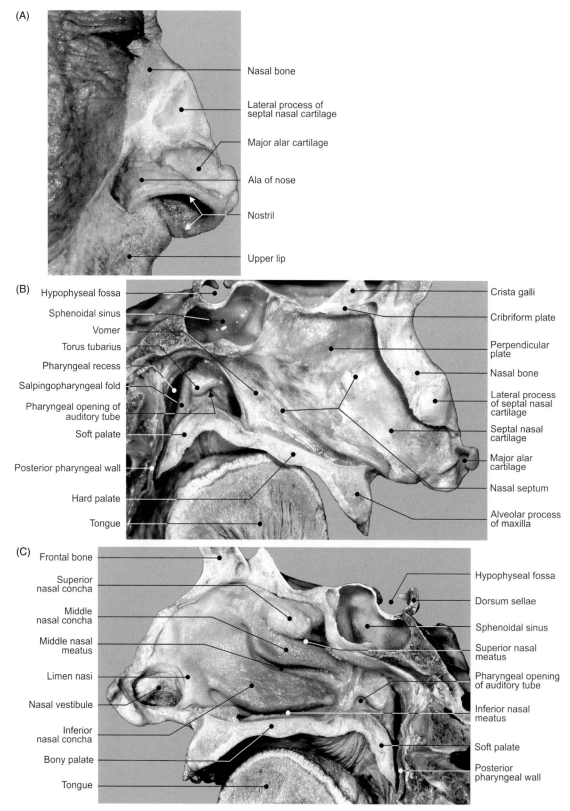

(A)

Nasal bone

Lateral process of
septal nasal cartilage

Major alar cartilage

Ala of nose

Nostril

Upper lip

(B)

Hypophyseal fossa

Sphenoidal sinus

Vomer

Torus tubarius

Pharyngeal recess

Salpingopharyngeal fold

Pharyngeal opening of
auditory tube

Soft palate

Posterior pharyngeal wall

Hard palate

Tongue

Crista galli

Cribriform plate

Perpendicular plate

Nasal bone

Lateral process
of septal nasal
cartilage

Septal nasal
cartilage

Major alar
cartilage

Nasal septum

Alveolar process
of maxilla

(C)

Frontal bone

Superior nasal concha

Middle nasal concha

Middle nasal meatus

Limen nasi

Nasal vestibule

Inferior nasal concha

Bony palate

Tongue

Hypophyseal fossa

Dorsum sellae

Sphenoidal sinus

Superior nasal meatus

Pharyngeal opening
of auditory tube

Inferior nasal meatus

Soft palate

Posterior
pharyngeal wall

FIGURE 6.38 The nasal and nasopharyngeal regions. (A) External nasal cartilages, (B) nasal septum, and (C) lateral nasal wall and nasopharynx.

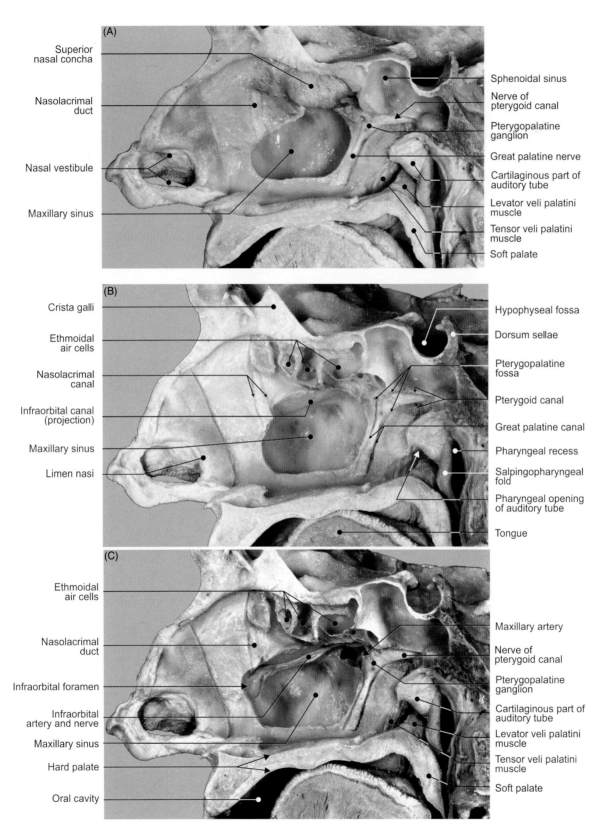

FIGURE 6.39 **Dissection from the nasal cavity of the lateral nasal and nasopharyngeal walls.** (A) First, (B) second, and (C) third layers.

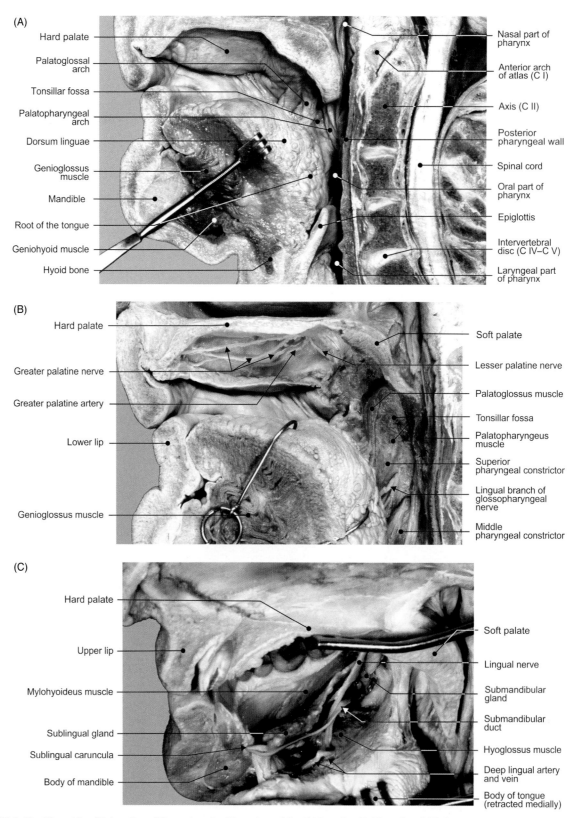

(A)

Hard palate
Palatoglossal arch
Tonsillar fossa
Palatopharyngeal arch
Dorsum linguae
Genioglossus muscle
Mandible
Root of the tongue
Geniohyoid muscle
Hyoid bone

Nasal part of pharynx
Anterior arch of atlas (C I)
Axis (C II)
Posterior pharyngeal wall
Spinal cord
Oral part of pharynx
Epiglottis
Intervertebral disc (C IV–C V)
Laryngeal part of pharynx

(B)

Hard palate
Greater palatine nerve
Greater palatine artery
Lower lip
Genioglossus muscle

Soft palate
Lesser palatine nerve
Palatoglossus muscle
Tonsillar fossa
Palatopharyngeus muscle
Superior pharyngeal constrictor
Lingual branch of glossopharyngeal nerve
Middle pharyngeal constrictor

(C)

Hard palate
Upper lip
Mylohyoideus muscle
Sublingual gland
Sublingual caruncula
Body of mandible

Soft palate
Lingual nerve
Submandibular gland
Submandibular duct
Hyoglossus muscle
Deep lingual artery and vein
Body of tongue (retracted medially)

FIGURE 6.40 The midsaqittal section of the oral cavity. Dissections of the (A) lateral wall, (B) roof, and (C) floor.

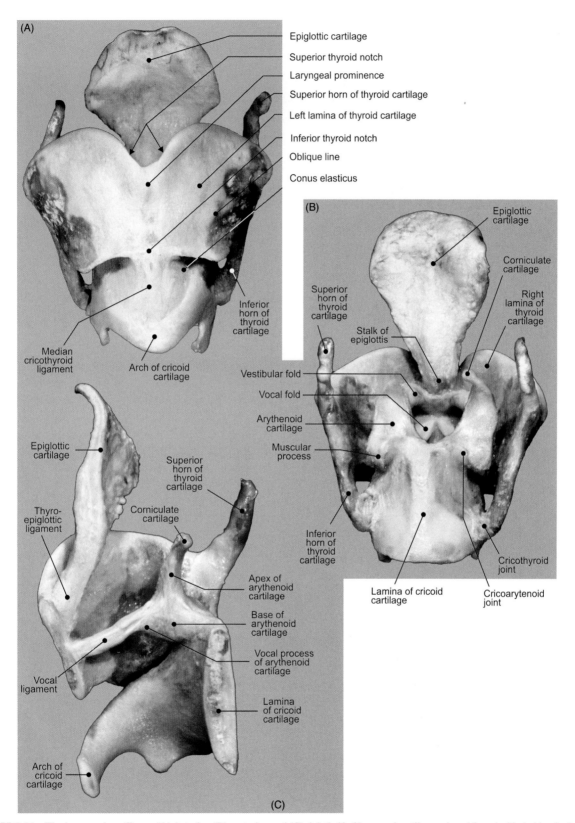

(A)

Epiglottic cartilage

Superior thyroid notch

Laryngeal prominence

Superior horn of thyroid cartilage

Left lamina of thyroid cartilage

Inferior thyroid notch

Oblique line

Conus elasticus

Inferior horn of thyroid cartilage

Median cricothyroid ligament

Arch of cricoid cartilage

(B)

Epiglottic cartilage

Corniculate cartilage

Right lamina of thyroid cartilage

Superior horn of thyroid cartilage

Stalk of epiglottis

Vestibular fold

Vocal fold

Arythenoid cartilage

Muscular process

Inferior horn of thyroid cartilage

Lamina of cricoid cartilage

Cricothyroid joint

Cricoarytenoid joint

Epiglottic cartilage

Thyro-epiglottic ligament

Superior horn of thyroid cartilage

Corniculate cartilage

Apex of arythenoid cartilage

Base of arythenoid cartilage

Vocal process of arythenoid cartilage

Lamina of cricoid cartilage

Vocal ligament

Arch of cricoid cartilage

(C)

FIGURE 6.41   The laryngeal cartilages. (A) Anterior, (B) posterior, and (C) right half of laryngeal cartilages viewed from inside (midsagittal section of laryngeal cartilages).

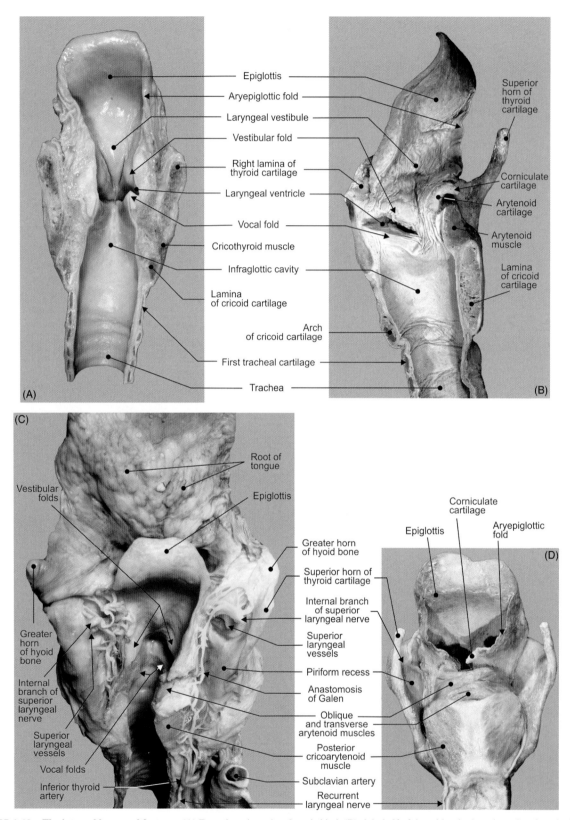

FIGURE 6.42 **The internal laryngeal features.** (A) Frontal section, view from behind; (B) right half of the midsagittal section, view from inside; (C) dissection of the piriform recess; and (D) dissection of the posterior laryngeal muscles.

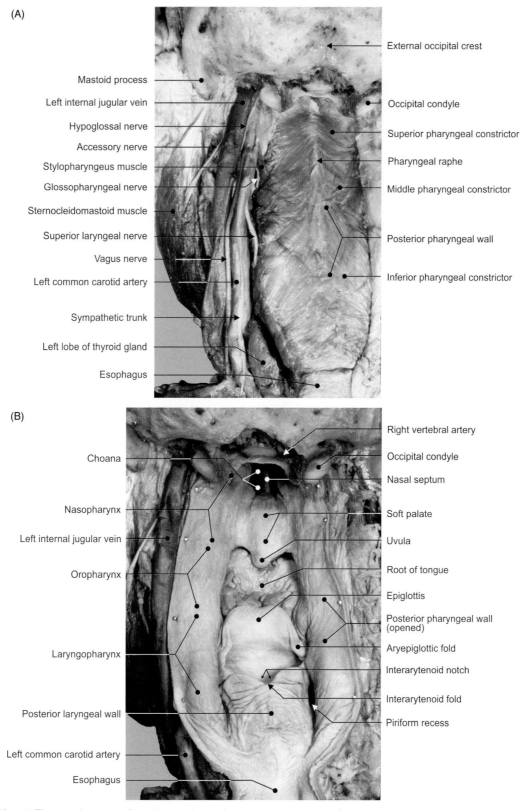

(A)

External occipital crest

Mastoid process

Left internal jugular vein — Occipital condyle

Hypoglossal nerve — Superior pharyngeal constrictor

Accessory nerve

Stylopharyngeus muscle — Pharyngeal raphe

Glossopharyngeal nerve — Middle pharyngeal constrictor

Sternocleidomastoid muscle

Superior laryngeal nerve

Vagus nerve — Posterior pharyngeal wall

Left common carotid artery — Inferior pharyngeal constrictor

Sympathetic trunk

Left lobe of thyroid gland

Esophagus

(B)

Right vertebral artery

Choana — Occipital condyle

Nasal septum

Nasopharynx — Soft palate

Left internal jugular vein — Uvula

Oropharynx — Root of tongue

Epiglottis

Posterior pharyngeal wall (opened)

Aryepiglottic fold

Laryngopharynx — Interarytenoid notch

Interarytenoid fold

Posterior laryngeal wall — Piriform recess

Left common carotid artery

Esophagus

FIGURE 6.43 (A) The posterior aspect of the pharyngeal constrictor muscles and content of the parapharyngeal space. (B) Reflection of the posterior pharyngeal wall, internal pharyngeal structures.

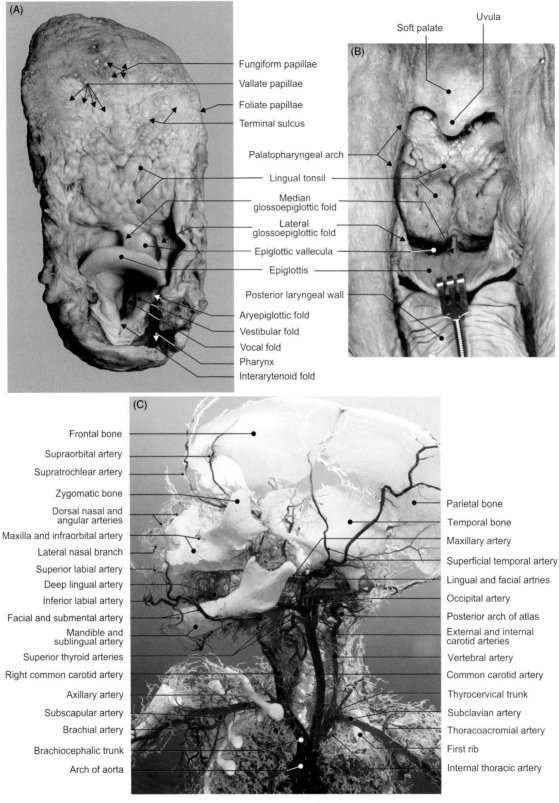

FIGURE 6.44 (A) Superior view of the tongue and larynx. (B) Root of tongue viewed from behind. (C) Fetal arterial corrosion cast of head and neck.

Chapter 7

Cranial Central Nervous System and Spinal Cord

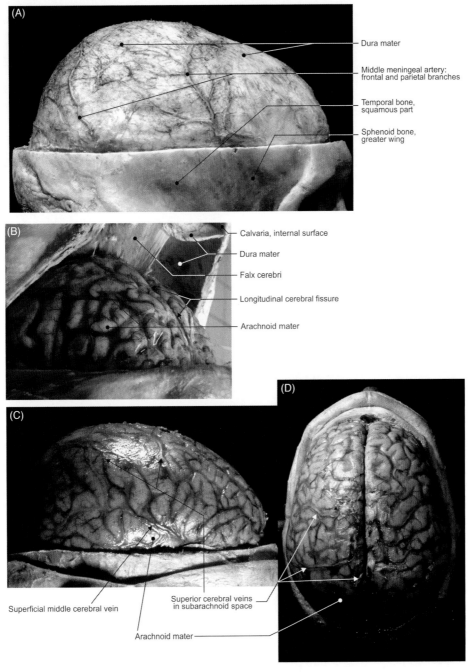

FIGURE 7.1 **Meninges.** (A, C–D) Formalin-fixed and (B) native specimens.

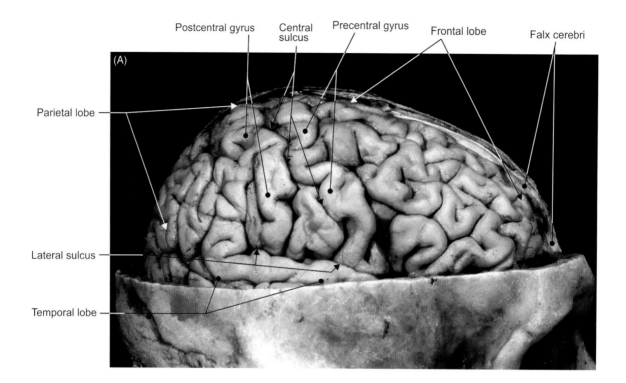

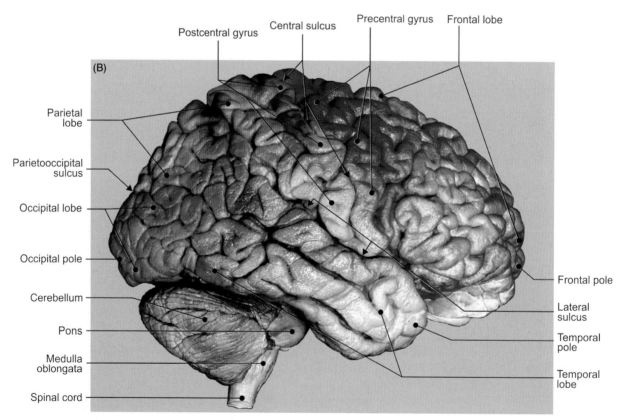

FIGURE 7.2 **The right cerebral hemisphere.** (A–B) Different views of the superolateral aspect.

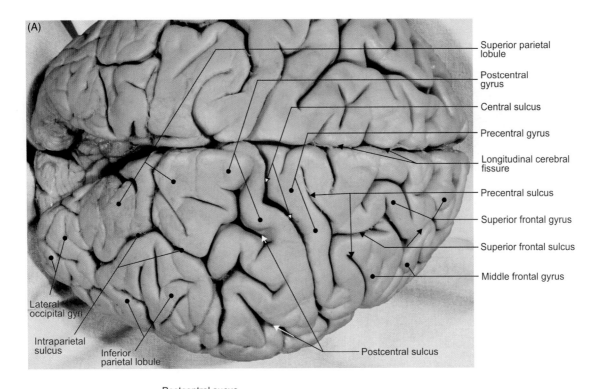

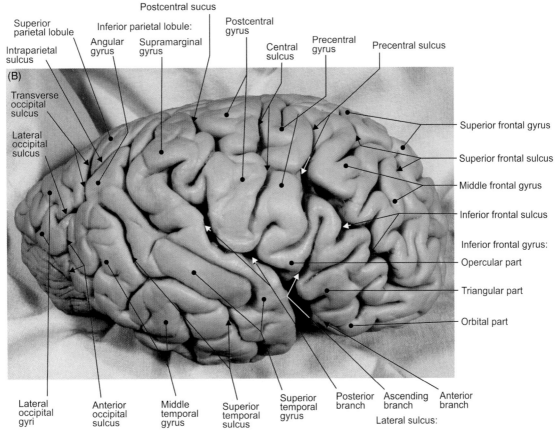

FIGURE 7.3   Superior aspect of (A) both hemispheres and (B) of the right hemisphere.

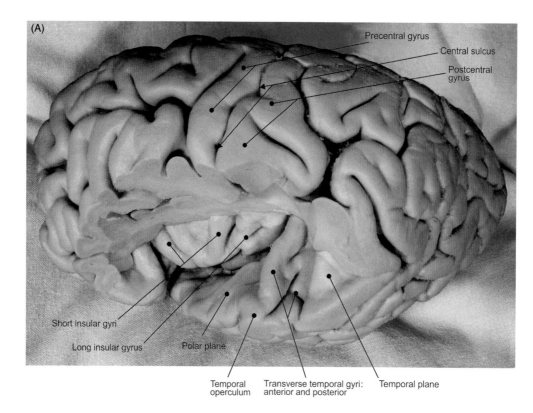

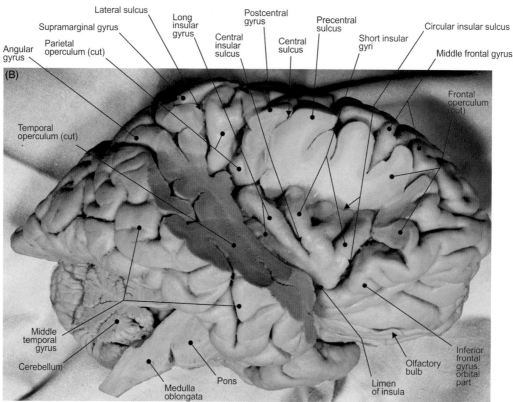

FIGURE 7.4 (A–B) Different views of the superolateral aspect of a hemisphere.

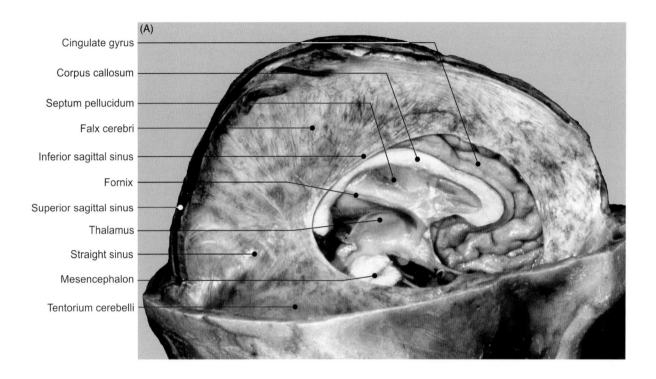

Cingulate gyrus
Corpus callosum
Septum pellucidum
Falx cerebri
Inferior sagittal sinus
Fornix
Superior sagittal sinus
Thalamus
Straight sinus
Mesencephalon
Tentorium cerebelli

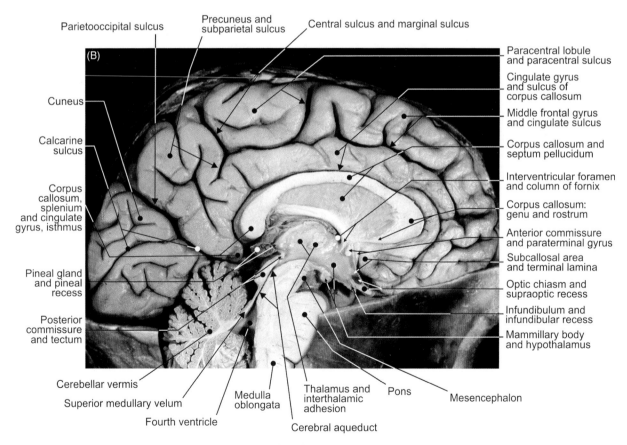

Parietooccipital sulcus
Precuneus and subparietal sulcus
Central sulcus and marginal sulcus
Cuneus
Calcarine sulcus
Corpus callosum, splenium and cingulate gyrus, isthmus
Pineal gland and pineal recess
Posterior commissure and tectum

Paracentral lobule and paracentral sulcus
Cingulate gyrus and sulcus of corpus callosum
Middle frontal gyrus and cingulate sulcus
Corpus callosum and septum pellucidum
Interventricular foramen and column of fornix
Corpus callosum: genu and rostrum
Anterior commissure and paraterminal gyrus
Subcallosal area and terminal lamina
Optic chiasm and supraoptic recess
Infundibulum and infundibular recess
Mammillary body and hypothalamus

Cerebellar vermis
Superior medullary velum
Fourth ventricle
Medulla oblongata
Thalamus and interthalamic adhesion
Cerebral aqueduct
Pons
Mesencephalon

FIGURE 7.5 (A) Falx cerebri and (B) the medial aspect of the left hemisphere.

(A)

(B)

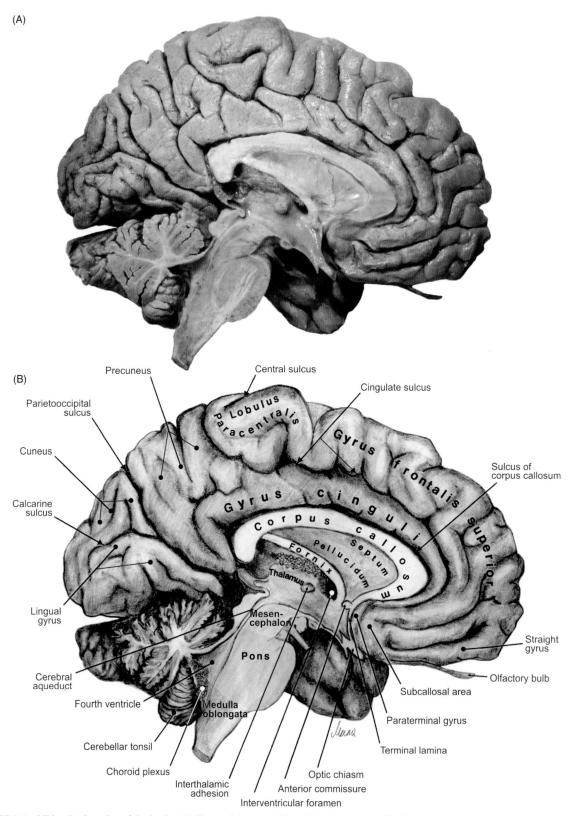

FIGURE 7.6 Midsagittal section of the brain. (A) The medial aspect of the left hemisphere and (B) illustration.

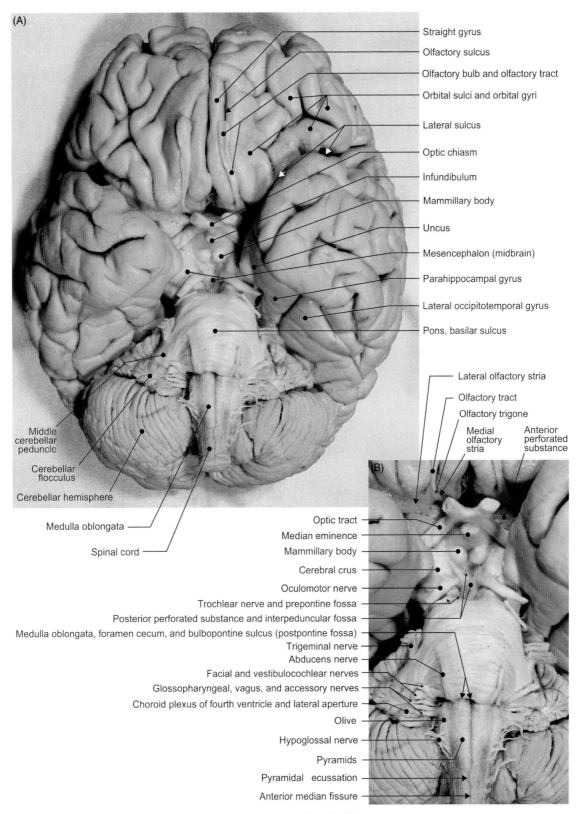

Straight gyrus

Olfactory sulcus

Olfactory bulb and olfactory tract

Orbital sulci and orbital gyri

Lateral sulcus

Optic chiasm

Infundibulum

Mammillary body

Uncus

Mesencephalon (midbrain)

Parahippocampal gyrus

Lateral occipitotemporal gyrus

Pons, basilar sulcus

Lateral olfactory stria

Olfactory tract

Olfactory trigone

Medial olfactory stria

Anterior perforated substance

Middle cerebellar peduncle

Cerebellar flocculus

Cerebellar hemisphere

Medulla oblongata

Spinal cord

Optic tract

Median eminence

Mammillary body

Cerebral crus

Oculomotor nerve

Trochlear nerve and prepontine fossa

Posterior perforated substance and interpeduncular fossa

Medulla oblongata, foramen cecum, and bulbopontine sulcus (postpontine fossa)

Trigeminal nerve

Abducens nerve

Facial and vestibulocochlear nerves

Glossopharyngeal, vagus, and accessory nerves

Choroid plexus of fourth ventricle and lateral aperture

Olive

Hypoglossal nerve

Pyramids

Pyramidal decussation

Anterior median fissure

FIGURE 7.7   Cerebrum, cerebellum, and brainstem. (A) Inferior aspect and (B) details.

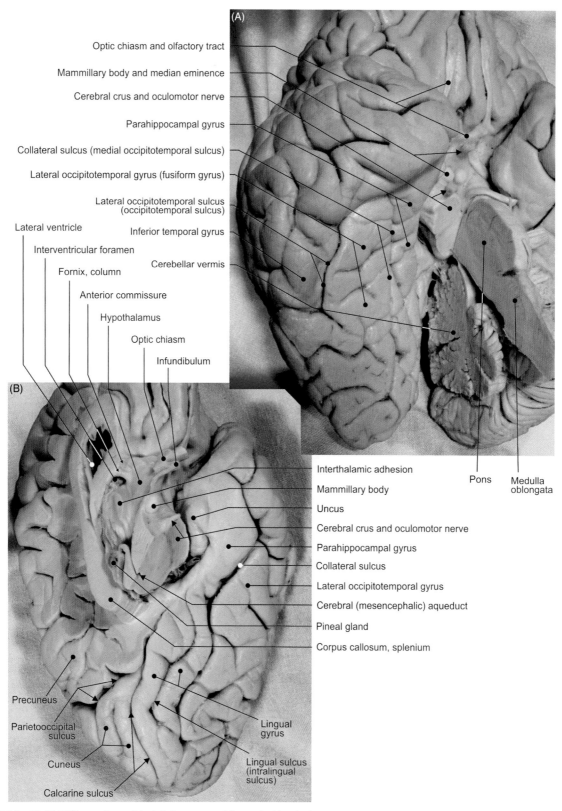

Optic chiasm and olfactory tract

Mammillary body and median eminence

Cerebral crus and oculomotor nerve

Parahippocampal gyrus

Collateral sulcus (medial occipitotemporal sulcus)

Lateral occipitotemporal gyrus (fusiform gyrus)

Lateral occipitotemporal sulcus (occipitotemporal sulcus)

Lateral ventricle

Interventricular foramen

Fornix, column

Anterior commissure

Hypothalamus

Optic chiasm

Infundibulum

Inferior temporal gyrus

Cerebellar vermis

Interthalamic adhesion

Mammillary body

Uncus

Cerebral crus and oculomotor nerve

Parahippocampal gyrus

Collateral sulcus

Lateral occipitotemporal gyrus

Cerebral (mesencephalic) aqueduct

Pineal gland

Corpus callosum, splenium

Pons

Medulla oblongata

Precuneus

Parietooccipital sulcus

Cuneus

Calcarine sulcus

Lingual gyrus

Lingual sulcus (intralingual sulcus)

FIGURE 7.8 (A–B) Different views of the inferior aspects of the cerebral hemispheres.

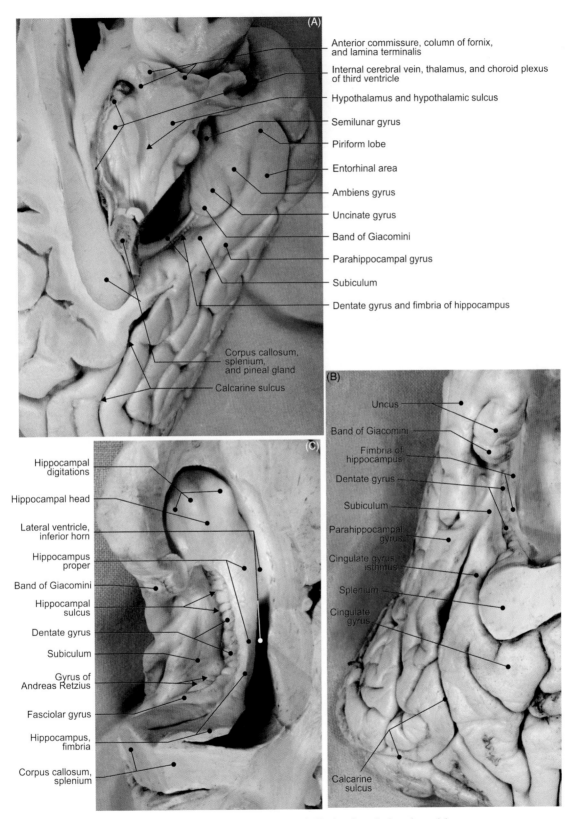

(A)

Anterior commissure, column of fornix, and lamina terminalis

Internal cerebral vein, thalamus, and choroid plexus of third ventricle

Hypothalamus and hypothalamic sulcus

Semilunar gyrus

Piriform lobe

Entorhinal area

Ambiens gyrus

Uncinate gyrus

Band of Giacomini

Parahippocampal gyrus

Subiculum

Dentate gyrus and fimbria of hippocampus

Corpus callosum, splenium, and pineal gland

Calcarine sulcus

(B)

Uncus

Band of Giacomini

Fimbria of hippocampus

Dentate gyrus

Subiculum

Parahippocampal gyrus

Cingulate gyrus, isthmus

Splenium

Cingulate gyrus

Calcarine sulcus

(C)

Hippocampal digitations

Hippocampal head

Lateral ventricle, inferior horn

Hippocampus proper

Band of Giacomini

Hippocampal sulcus

Dentate gyrus

Subiculum

Gyrus of Andreas Retzius

Fasciolar gyrus

Hippocampus, fimbria

Corpus callosum, splenium

FIGURE 7.9   The hippocampus. (A) Inferomedial view, (B) medial view, and (C) view from the lateral ventricle.

(A)

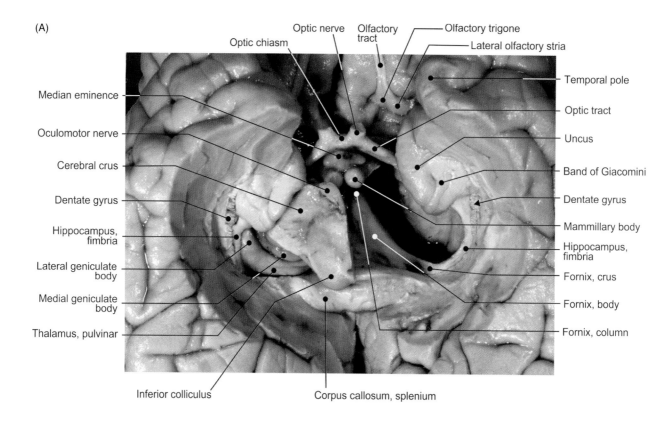

Optic nerve
Optic chiasm
Olfactory tract
Olfactory trigone
Lateral olfactory stria

Median eminence
Oculomotor nerve
Cerebral crus
Dentate gyrus
Hippocampus, fimbria
Lateral geniculate body
Medial geniculate body
Thalamus, pulvinar

Temporal pole
Optic tract
Uncus
Band of Giacomini
Dentate gyrus
Mammillary body
Hippocampus, fimbria
Fornix, crus
Fornix, body
Fornix, column

Inferior colliculus
Corpus callosum, splenium

(B)

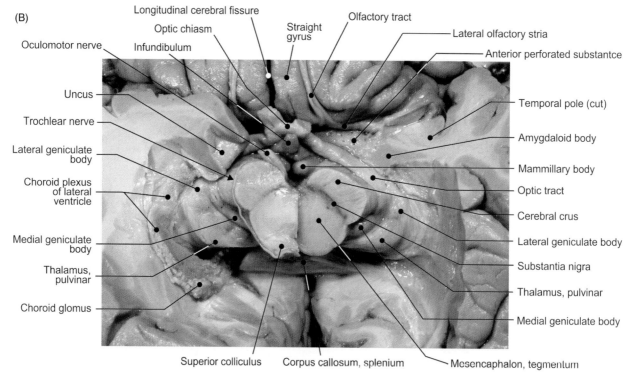

Longitudinal cerebral fissure
Optic chiasm
Infundibulum
Straight gyrus
Olfactory tract
Lateral olfactory stria
Anterior perforated substantce

Oculomotor nerve
Uncus
Trochlear nerve
Lateral geniculate body
Choroid plexus of lateral ventricle
Medial geniculate body
Thalamus, pulvinar
Choroid glomus

Temporal pole (cut)
Amygdaloid body
Mammillary body
Optic tract
Cerebral crus
Lateral geniculate body
Substantia nigra
Thalamus, pulvinar
Medial geniculate body

Superior colliculus
Corpus callosum, splenium
Mesencaphalon, tegmentum

FIGURE 7.10 The inferior aspect of encephalon. (A) Removed: both parahippocampal fimbriae, caudal segment of the brainstem, and thalamus of the left side. (B) Removed: both parahippocampal fimbriae and caudal segment of the brainstem.

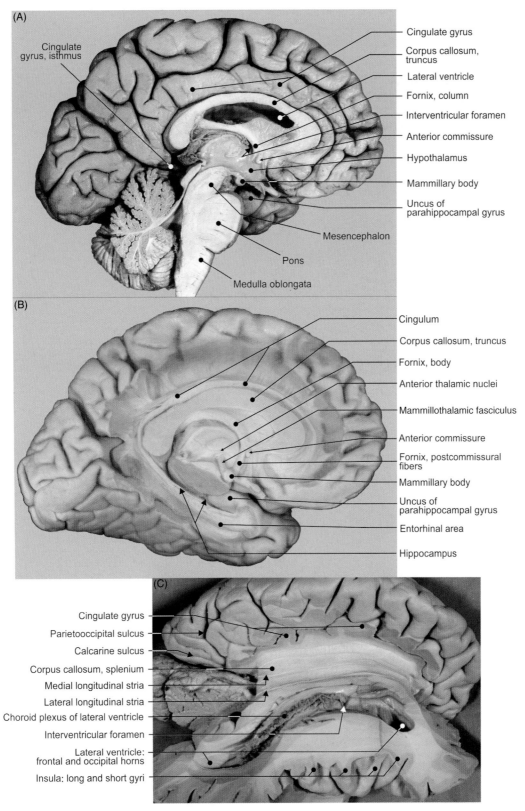

FIGURE 7.11   The limbic lobe. (A) Medial aspect, (B) dissection, and (C) corpus callosum, view from above.

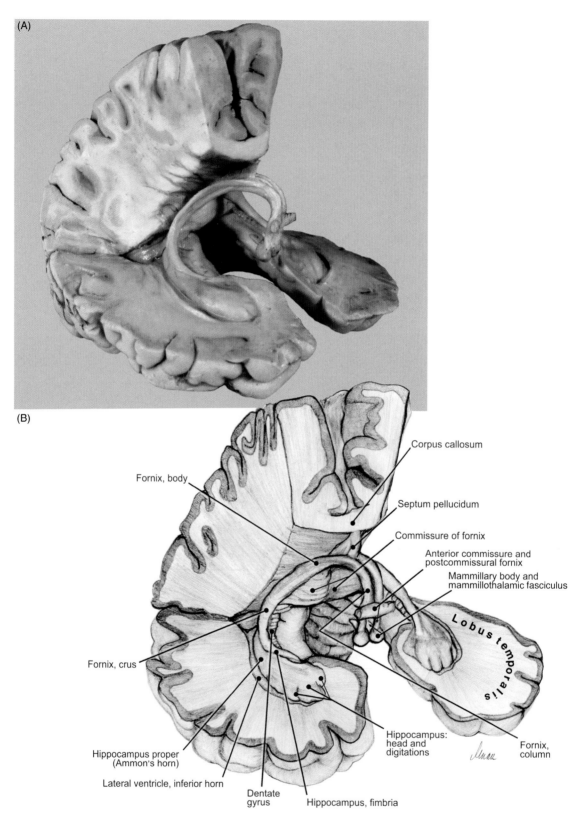

(A)

(B)

Corpus callosum

Septum pellucidum

Commissure of fornix

Anterior commissure and
postcommissural fornix

Mammillary body and
mammillothalamic fasciculus

Fornix, body

Lobus temporalis

Fornix, crus

Hippocampus:
head and
digitations

Fornix,
column

Hippocampus proper
(Ammon's horn)

Lateral ventricle, inferior horn

Dentate
gyrus

Hippocampus, fimbria

FIGURE 7.12 **The fornix and hippocampus.** (A) Dissection and (B) illustration.

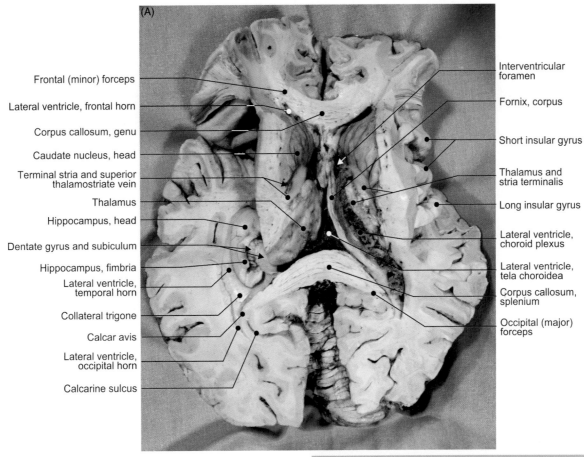

Frontal (minor) forceps

Lateral ventricle, frontal horn

Corpus callosum, genu

Caudate nucleus, head

Terminal stria and superior thalamostriate vein

Thalamus

Hippocampus, head

Dentate gyrus and subiculum

Hippocampus, fimbria

Lateral ventricle, temporal horn

Collateral trigone

Calcar avis

Lateral ventricle, occipital horn

Calcarine sulcus

Interventricular foramen

Fornix, corpus

Short insular gyrus

Thalamus and stria terminalis

Long insular gyrus

Lateral ventricle, choroid plexus

Lateral ventricle, tela choroidea

Corpus callosum, splenium

Occipital (major) forceps

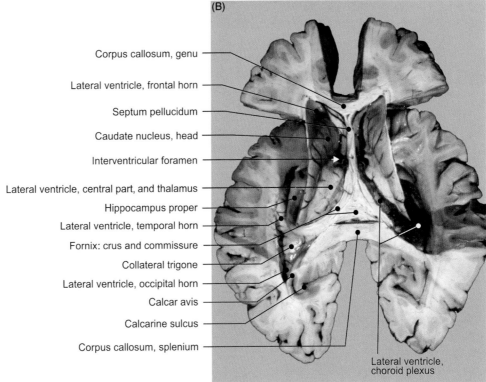

Corpus callosum, genu

Lateral ventricle, frontal horn

Septum pellucidum

Caudate nucleus, head

Interventricular foramen

Lateral ventricle, central part, and thalamus

Hippocampus proper

Lateral ventricle, temporal horn

Fornix: crus and commissure

Collateral trigone

Lateral ventricle, occipital horn

Calcar avis

Calcarine sulcus

Corpus callosum, splenium

Lateral ventricle, choroid plexus

FIGURE 7.13 Lateral ventricles. (A–B) Different views from above.

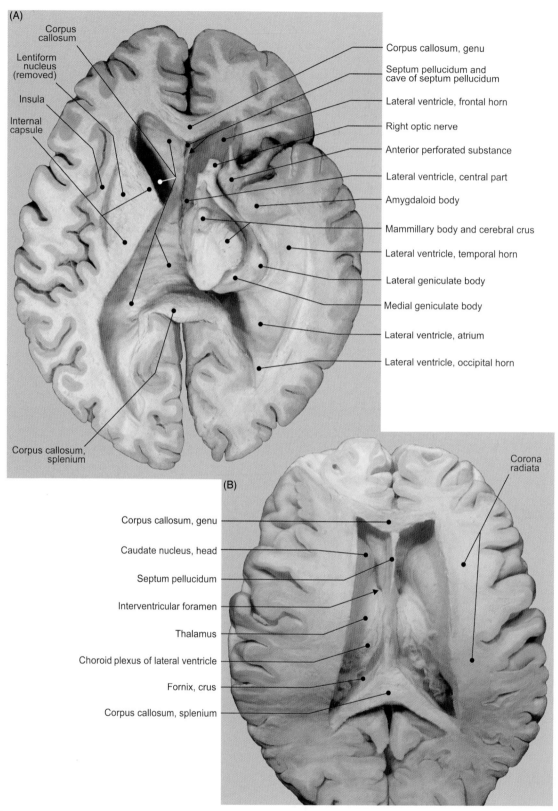

FIGURE 7.14 Dissected lateral ventricles. Views from (A) below and (B) from above.

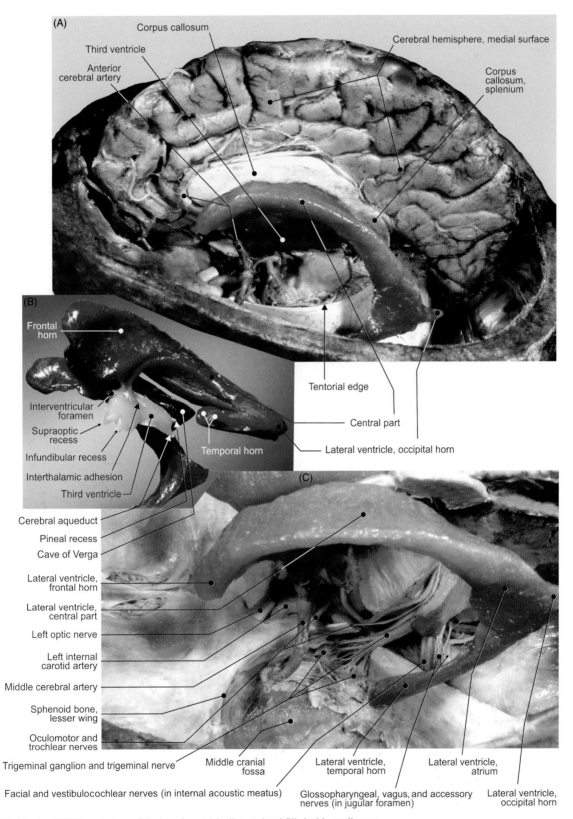

(A)

Corpus callosum

Third ventricle

Anterior
cerebral artery

Cerebral hemisphere, medial surface

Corpus
callosum,
splenium

(B)

Frontal
horn

Interventricular
foramen

Supraoptic
recess

Infundibular recess

Interthalamic adhesion

Third ventricle

Temporal horn

Tentorial edge

Central part

Lateral ventricle, occipital horn

(C)

Cerebral aqueduct

Pineal recess

Cave of Verga

Lateral ventricle,
frontal horn

Lateral ventricle,
central part

Left optic nerve

Left internal
carotid artery

Middle cerebral artery

Sphenoid bone,
lesser wing

Oculomotor and
trochlear nerves

Trigeminal ganglion and trigeminal nerve

Facial and vestibulocochlear nerves (in internal acoustic meatus)

Middle cranial
fossa

Lateral ventricle,
temporal horn

Glossopharyngeal, vagus, and accessory
nerves (in jugular foramen)

Lateral ventricle,
atrium

Lateral ventricle,
occipital horn

FIGURE 7.15   (A–C) Different views of the lateral ventricle dissected and filled with acrylic cast.

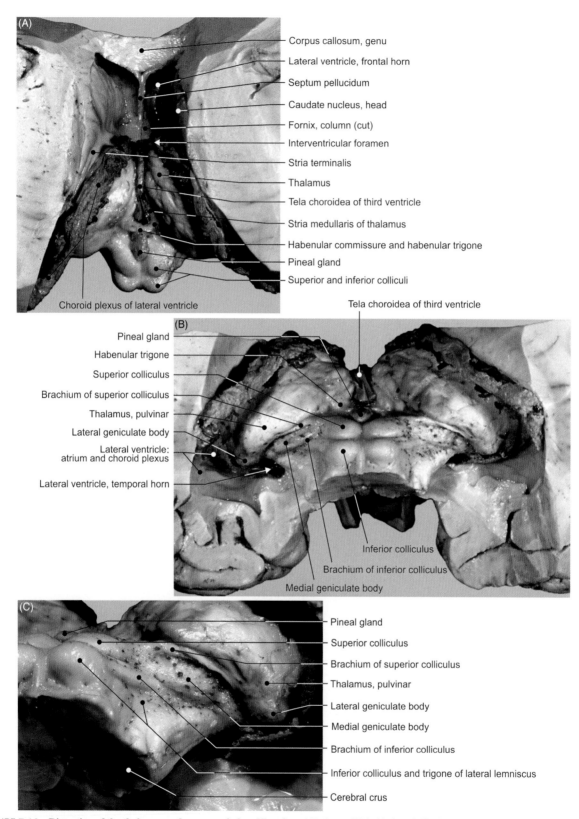

(A)

Corpus callosum, genu

Lateral ventricle, frontal horn

Septum pellucidum

Caudate nucleus, head

Fornix, column (cut)

Interventricular foramen

Stria terminalis

Thalamus

Tela choroidea of third ventricle

Stria medullaris of thalamus

Habenular commissure and habenular trigone

Pineal gland

Superior and inferior colliculi

Choroid plexus of lateral ventricle

Tela choroidea of third ventricle

(B)

Pineal gland

Habenular trigone

Superior colliculus

Brachium of superior colliculus

Thalamus, pulvinar

Lateral geniculate body

Lateral ventricle: atrium and choroid plexus

Lateral ventricle, temporal horn

Inferior colliculus

Brachium of inferior colliculus

Medial geniculate body

(C)

Pineal gland

Superior colliculus

Brachium of superior colliculus

Thalamus, pulvinar

Lateral geniculate body

Medial geniculate body

Brachium of inferior colliculus

Inferior colliculus and trigone of lateral lemniscus

Cerebral crus

FIGURE 7.16   Dissection of the thalamus and mesencephalon. View from (A) above, (B) behind, and (C) the lateral side.

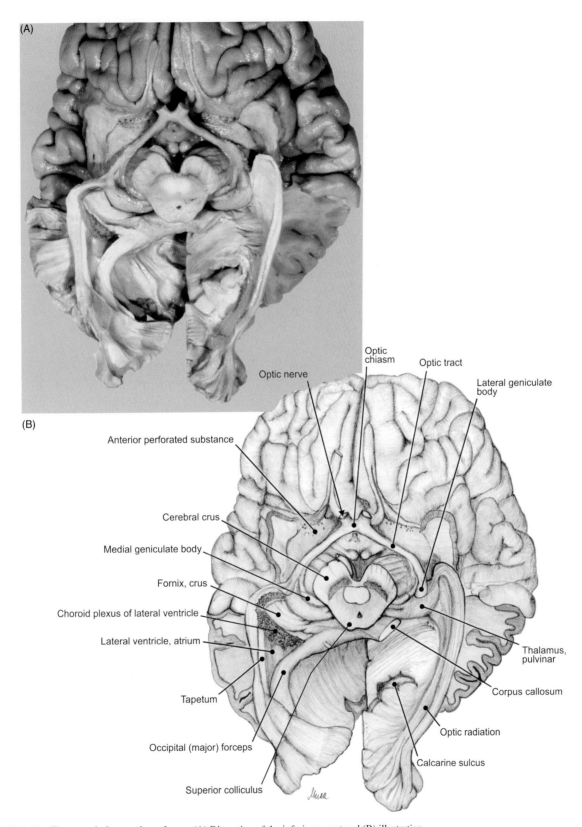

FIGURE 7.17   The encephalon, optic pathway. (A) Dissection of the inferior aspect and (B) illustration.

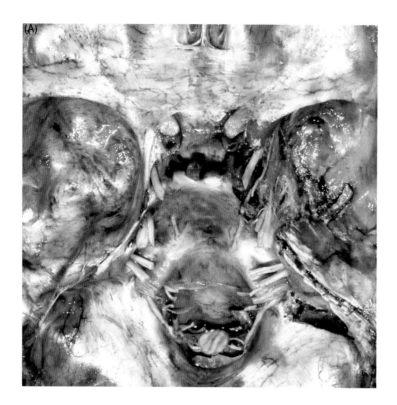

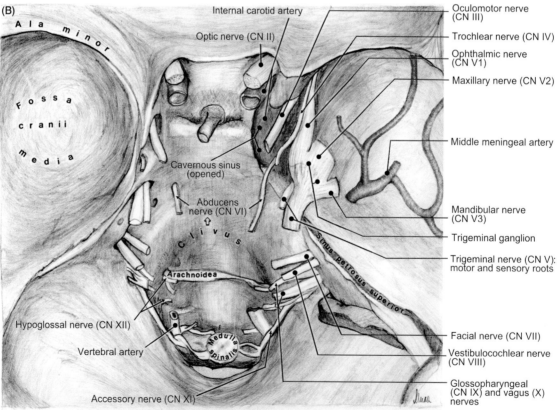

FIGURE 7.18 The base of the skull with passage of cranial nerves. (A) Dura mater removed from the right half of the cranial base and (B) illustration.

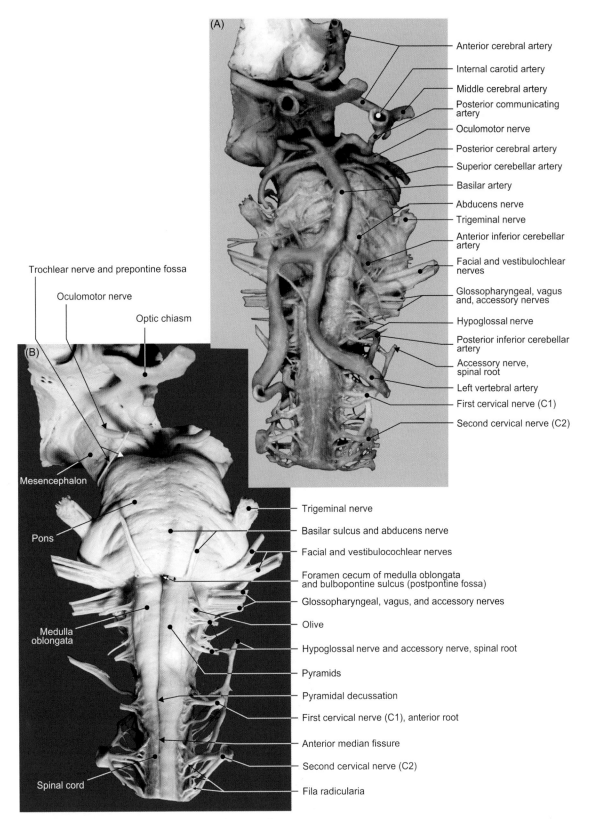

FIGURE 7.19 **The brainstem.** (A–B) Different views of the ventral aspect.

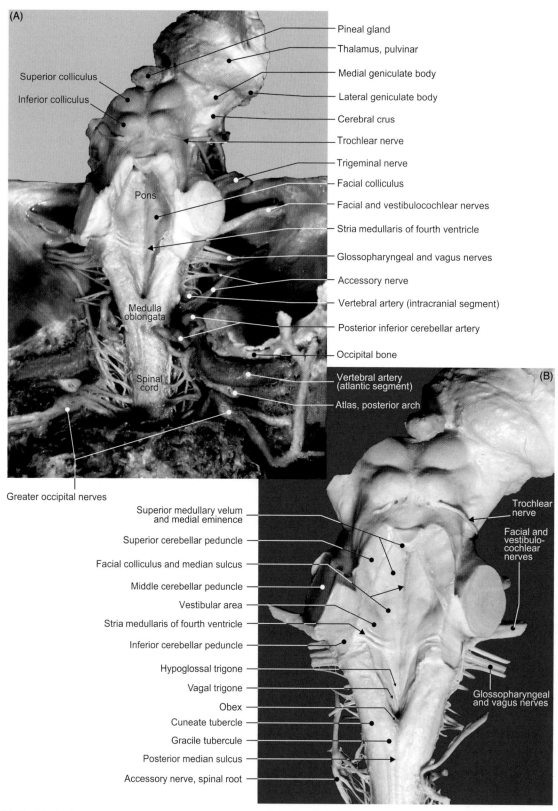

(A)

Pineal gland

Thalamus, pulvinar

Superior colliculus

Medial geniculate body

Inferior colliculus

Lateral geniculate body

Cerebral crus

Trochlear nerve

Trigeminal nerve

Facial colliculus

Pons

Facial and vestibulocochlear nerves

Stria medullaris of fourth ventricle

Glossopharyngeal and vagus nerves

Accessory nerve

Vertebral artery (intracranial segment)

Medulla
oblongata

Posterior inferior cerebellar artery

Occipital bone

(B)

Spinal
cord

Vertebral artery
(atlantic segment)

Atlas, posterior arch

Greater occipital nerves

Superior medullary velum
and medial eminence

Trochlear
nerve

Superior cerebellar peduncle

Facial and
vestibulo-
cochlear
nerves

Facial colliculus and median sulcus

Middle cerebellar peduncle

Vestibular area

Stria medullaris of fourth ventricle

Inferior cerebellar peduncle

Hypoglossal trigone

Vagal trigone

Obex

Glossopharyngeal
and vagus nerves

Cuneate tubercle

Gracile tubercule

Posterior median sulcus

Accessory nerve, spinal root

FIGURE 7.20   The brainstem. (A–B) Different views of the dorsal aspect.

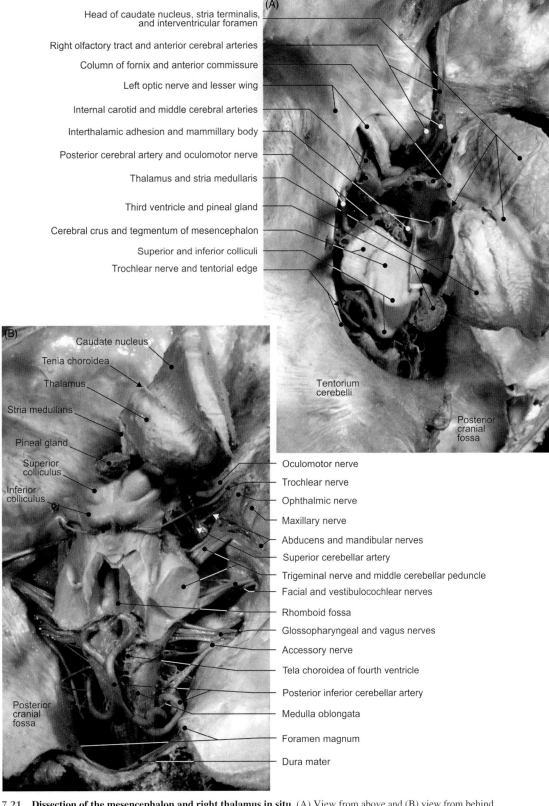

Head of caudate nucleus, stria terminalis, and interventricular foramen

Right olfactory tract and anterior cerebral arteries

Column of fornix and anterior commissure

Left optic nerve and lesser wing

Internal carotid and middle cerebral arteries

Interthalamic adhesion and mammillary body

Posterior cerebral artery and oculomotor nerve

Thalamus and stria medullaris

Third ventricle and pineal gland

Cerebral crus and tegmentum of mesencephalon

Superior and inferior colliculi

Trochlear nerve and tentorial edge

Tentorium cerebelli

Posterior cranial fossa

(B)

Caudate nucleus

Tenia choroidea

Thalamus

Stria medullaris

Pineal gland

Superior colliculus

Inferior colliculus

Posterior cranial fossa

Oculomotor nerve

Trochlear nerve

Ophthalmic nerve

Maxillary nerve

Abducens and mandibular nerves

Superior cerebellar artery

Trigeminal nerve and middle cerebellar peduncle

Facial and vestibulocochlear nerves

Rhomboid fossa

Glossopharyngeal and vagus nerves

Accessory nerve

Tela choroidea of fourth ventricle

Posterior inferior cerebellar artery

Medulla oblongata

Foramen magnum

Dura mater

FIGURE 7.21 **Dissection of the mesencephalon and right thalamus in situ.** (A) View from above and (B) view from behind.

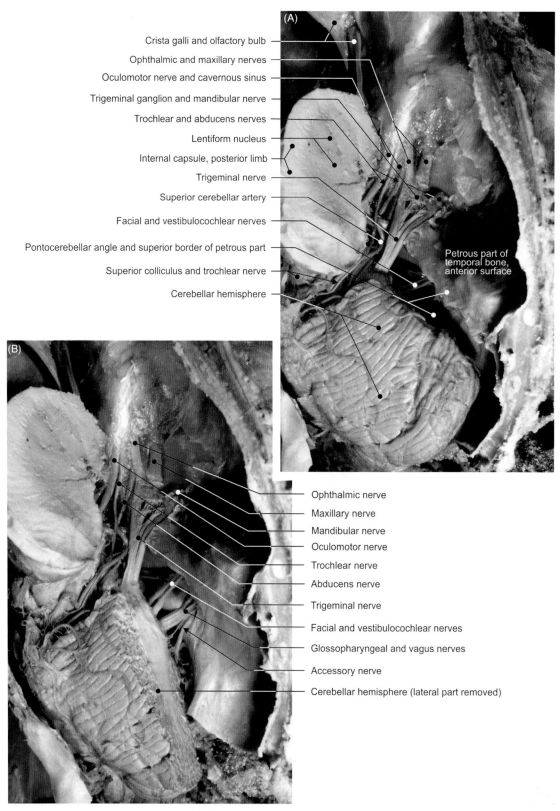

Crista galli and olfactory bulb
Ophthalmic and maxillary nerves
Oculomotor nerve and cavernous sinus
Trigeminal ganglion and mandibular nerve
Trochlear and abducens nerves
Lentiform nucleus
Internal capsule, posterior limb
Trigeminal nerve
Superior cerebellar artery
Facial and vestibulocochlear nerves
Pontocerebellar angle and superior border of petrous part
Superior colliculus and trochlear nerve
Cerebellar hemisphere

Petrous part of temporal bone, anterior surface

Ophthalmic nerve
Maxillary nerve
Mandibular nerve
Oculomotor nerve
Trochlear nerve
Abducens nerve
Trigeminal nerve
Facial and vestibulocochlear nerves
Glossopharyngeal and vagus nerves
Accessory nerve
Cerebellar hemisphere (lateral part removed)

FIGURE 7.22 (A–B) Different views of the dissection of the cavernous sinus and pontocerebellar angle.

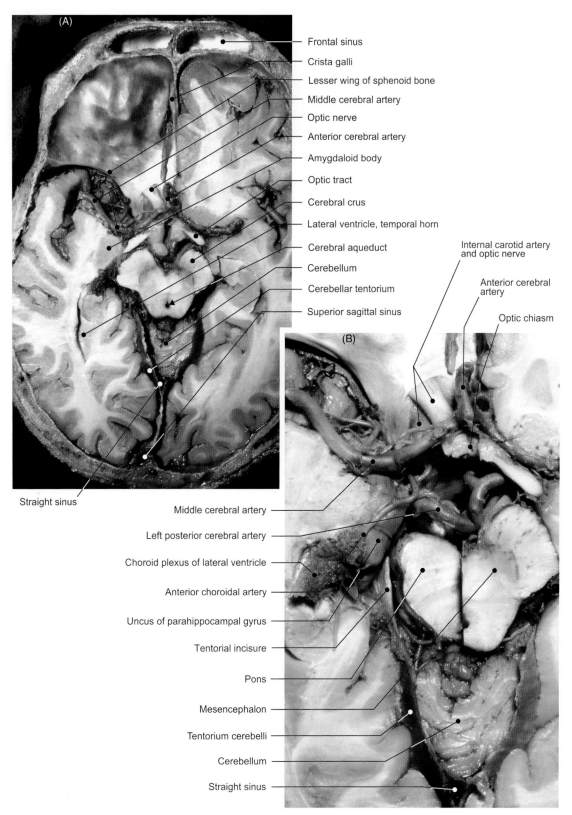

Frontal sinus

Crista galli

Lesser wing of sphenoid bone

Middle cerebral artery

Optic nerve

Anterior cerebral artery

Amygdaloid body

Optic tract

Cerebral crus

Lateral ventricle, temporal horn

Cerebral aqueduct

Cerebellum

Cerebellar tentorium

Superior sagittal sinus

Internal carotid artery and optic nerve

Anterior cerebral artery

Optic chiasm

Straight sinus

Middle cerebral artery

Left posterior cerebral artery

Choroid plexus of lateral ventricle

Anterior choroidal artery

Uncus of parahippocampal gyrus

Tentorial incisure

Pons

Mesencephalon

Tentorium cerebelli

Cerebellum

Straight sinus

FIGURE 7.23   (A–B) Transverse sections of the head: dissections.

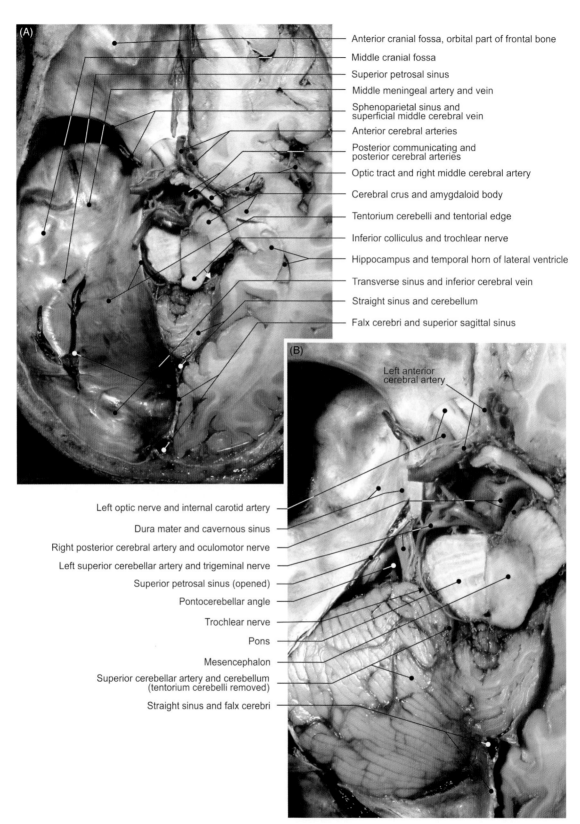

(A)

Anterior cranial fossa, orbital part of frontal bone
Middle cranial fossa
Superior petrosal sinus
Middle meningeal artery and vein
Sphenoparietal sinus and superficial middle cerebral vein
Anterior cerebral arteries
Posterior communicating and posterior cerebral arteries
Optic tract and right middle cerebral artery
Cerebral crus and amygdaloid body
Tentorium cerebelli and tentorial edge
Inferior colliculus and trochlear nerve
Hippocampus and temporal horn of lateral ventricle
Transverse sinus and inferior cerebral vein
Straight sinus and cerebellum
Falx cerebri and superior sagittal sinus

(B)

Left anterior cerebral artery

Left optic nerve and internal carotid artery
Dura mater and cavernous sinus
Right posterior cerebral artery and oculomotor nerve
Left superior cerebellar artery and trigeminal nerve
Superior petrosal sinus (opened)
Pontocerebellar angle
Trochlear nerve
Pons
Mesencephalon
Superior cerebellar artery and cerebellum (tentorium cerebelli removed)
Straight sinus and falx cerebri

FIGURE 7.24 (A–B) Transverse sections of the head: dissections.

(A)

Tentorium cerebelli

Superior sagittal sinus

Falx cerebri

Occipital pole

Confluence of sinuses

Transverse sinus

Cerebellar hemisphere

Arachnoid mater

Cerebellomedullary cistern

Dura mater

Foramen magnum

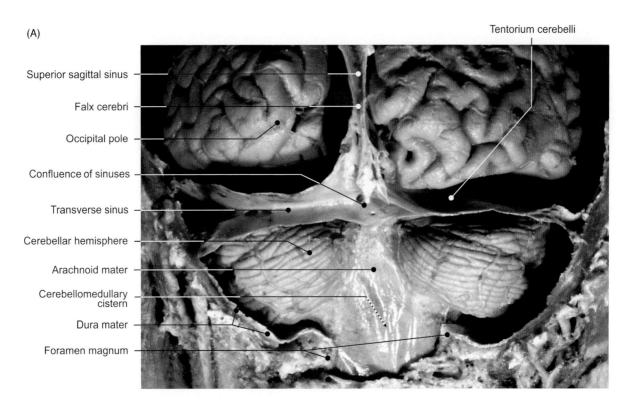

(B)

Falx cerebri

Dura mater

Superior cerebral veins (openings)

Tentorial incisure

Tentorium cerebelli

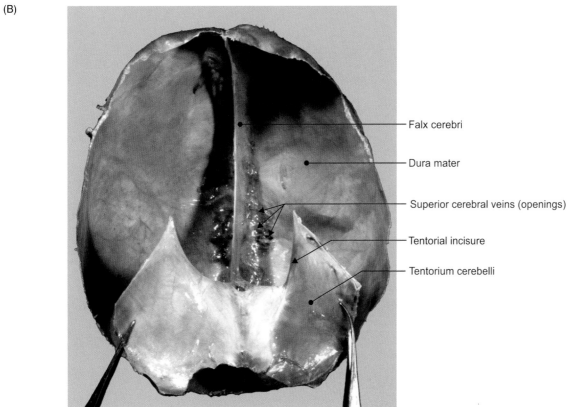

FIGURE 7.25 (A) The telencephalon and cerebellum in situ: view from behind. (B) Isolated cranial dura mater: view from below.

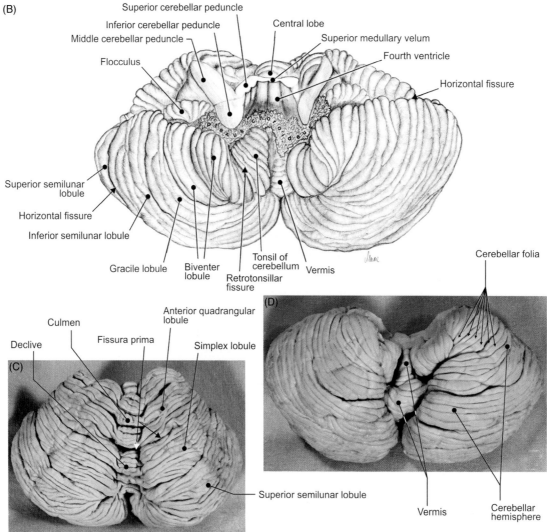

FIGURE 7.26 **The cerebellum.** (A) View from the front, (C) view from above, and (D) view from below. (B) Illustration of the cerebellum.

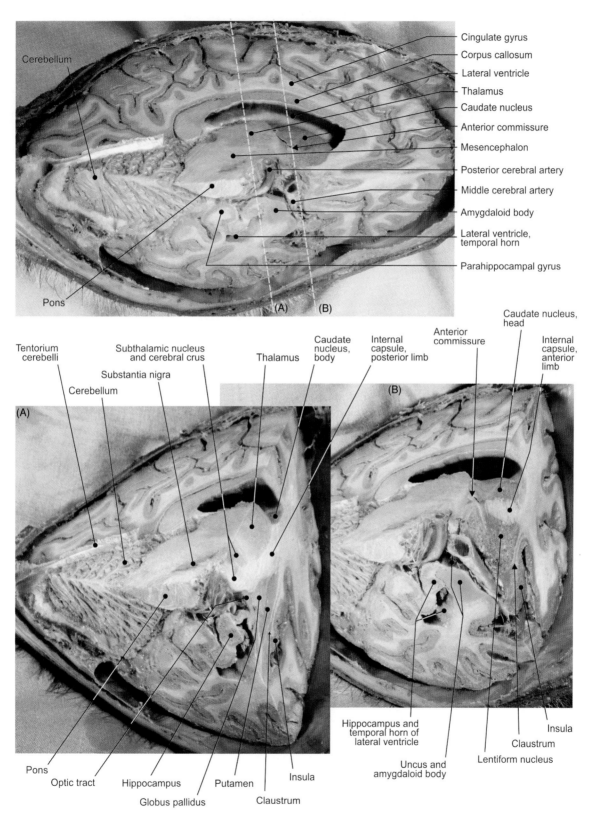

Cerebellum

Cingulate gyrus
Corpus callosum
Lateral ventricle
Thalamus
Caudate nucleus
Anterior commissure
Mesencephalon
Posterior cerebral artery
Middle cerebral artery
Amygdaloid body
Lateral ventricle, temporal horn
Parahippocampal gyrus

Pons

(A)   (B)

Tentorium cerebelli

Subthalamic nucleus and cerebral crus

Substantia nigra

Cerebellum

(A)

Caudate nucleus, body

Internal capsule, posterior limb

Thalamus

Anterior commissure

Caudate nucleus, head

Internal capsule, anterior limb

(B)

Pons

Optic tract

Hippocampus

Putamen

Insula

Globus pallidus

Claustrum

Hippocampus and temporal horn of lateral ventricle

Uncus and amygdaloid body

Lentiform nucleus

Claustrum

Insula

FIGURE 7.27   **Different views of the sections of the cerebral hemisphere.** (A) and (B) are the frontal sections.

(A)

Interventricular foramen
Caudate nucleus, head
Internal capsule, genu
Putamen
Lateral cerebral fossa
Insula
Claustrum
Globus pallidus
Oculomotor nerve
Pons, basilar part
Clivus
Internal carotid artery
Vertebral veins
Styloid process
Vertebral artery
Internal jugular vein

Corpus callosum, truncus
Lateral ventricle
Septum peliucidum
Choroid plexus
Fornix, column
Third ventricle
Optic tract
Amygdaloid body
Lateral ventricle
Basilar artery
Trigeminal nerve
Abducens nerve
Internal jugular vein, superior bulb
Occipital condyle
Vertebral artery
Axis, dens
Atlas, lateral mass

(B)

Thalamus
Putamen
Claustrum
Red nucleus
Subthalamic nucleus
Caudate nucleus, tail
Substantia nigra
Trochlear nerve
Facial, intermedius, and vestibulocochlear nerves
Glossopharyngeal and vagus nerves
Internal jugular vein, superior bulb
Atlantooccipital joint
Vertebral artery
Atlas, posterior arch
Axis

Pons

Corpus callosum, truncus
Caudate nucleus, body
Lateral ventricle
Fornix
Internal capsule, posterior limb
Third ventricle
Posterior cerebral artery
Lateral ventricle
Hippocampus
Tentorium cerebelli
Temporal bone, petrous part
Medulla oblongata
Sigmoid sinus
Vertebral artery
Spinal cord
Spinal nerves, roots
Obliquus capitis inferior muscle
Spinal dura mater

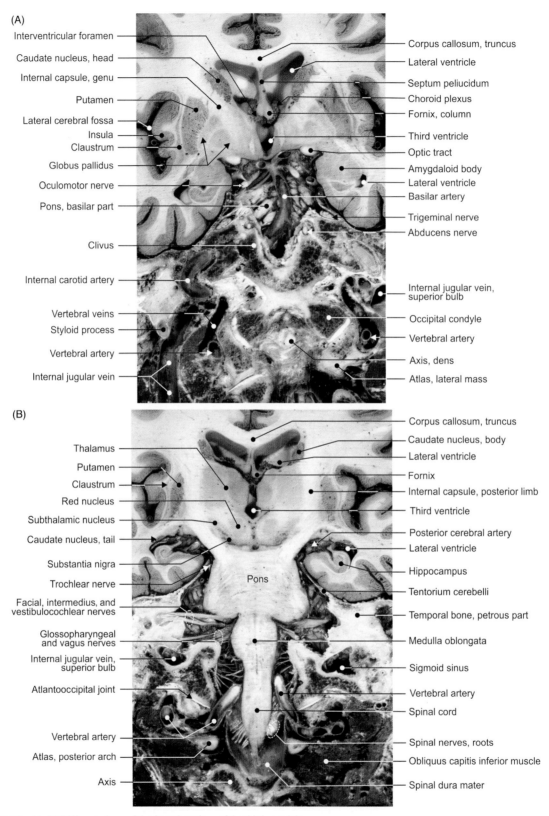

FIGURE 7.28 (A–B) Different views of the frontal sections of the third ventricle.

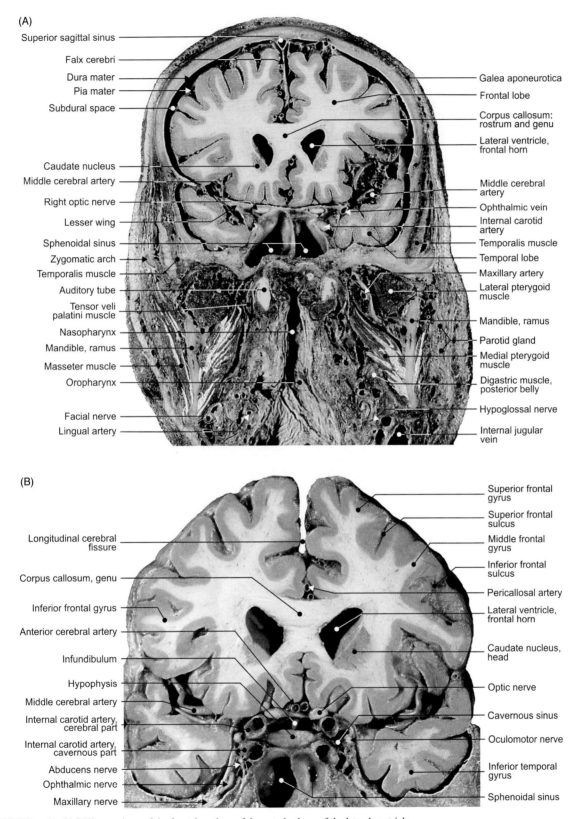

FIGURE 7.29   (A–B) Different views of the frontal sections of the anterior horn of the lateral ventricle.

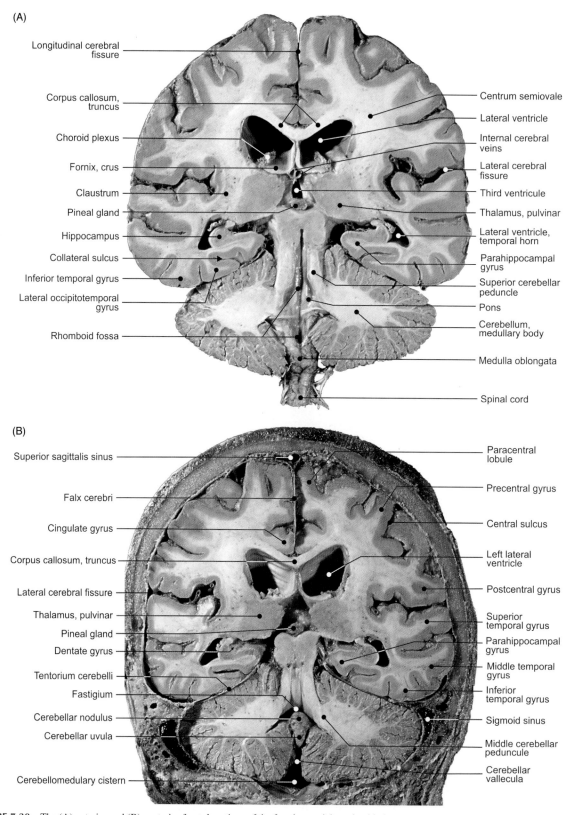

FIGURE 7.30 The (A) anterior and (B) posterior frontal sections of the fourth ventricle and epithalamus.

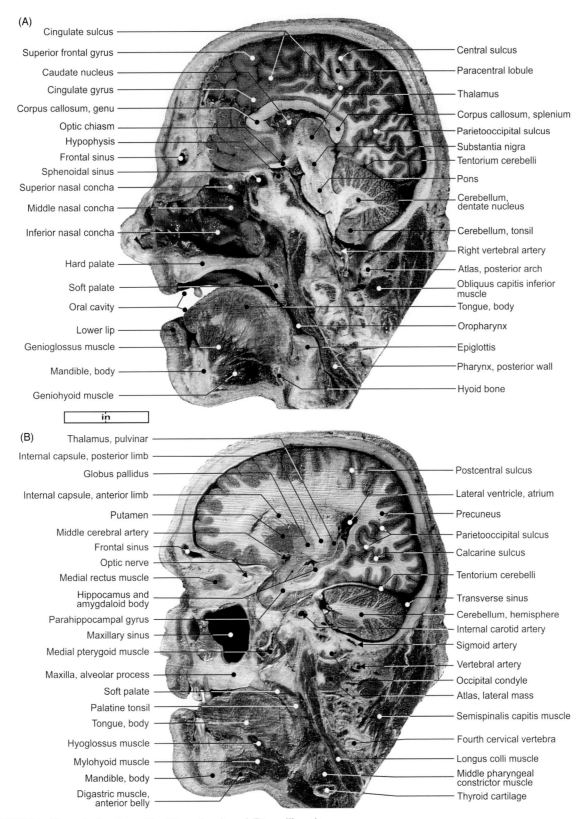

(A)

Cingulate sulcus

Superior frontal gyrus

Caudate nucleus

Cingulate gyrus

Corpus callosum, genu

Optic chiasm

Hypophysis

Frontal sinus

Sphenoidal sinus

Superior nasal concha

Middle nasal concha

Inferior nasal concha

Hard palate

Soft palate

Oral cavity

Lower lip

Genioglossus muscle

Mandible, body

Geniohyoid muscle

Central sulcus

Paracentral lobule

Thalamus

Corpus callosum, splenium

Parietooccipital sulcus

Substantia nigra

Tentorium cerebelli

Pons

Cerebellum, dentate nucleus

Cerebellum, tonsil

Right vertebral artery

Atlas, posterior arch

Obliquus capitis inferior muscle

Tongue, body

Oropharynx

Epiglottis

Pharynx, posterior wall

Hyoid bone

in

(B)

Thalamus, pulvinar

Internal capsule, posterior limb

Globus pallidus

Internal capsule, anterior limb

Putamen

Middle cerebral artery

Frontal sinus

Optic nerve

Medial rectus muscle

Hippocamus and amygdaloid body

Parahippocampal gyrus

Maxillary sinus

Medial pterygoid muscle

Maxilla, alveolar process

Soft palate

Palatine tonsil

Tongue, body

Hyoglossus muscle

Mylohyoid muscle

Mandible, body

Digastric muscle, anterior belly

Postcentral sulcus

Lateral ventricle, atrium

Precuneus

Parietooccipital sulcus

Calcarine sulcus

Tentorium cerebelli

Transverse sinus

Cerebellum, hemisphere

Internal carotid artery

Sigmoid artery

Vertebral artery

Occipital condyle

Atlas, lateral mass

Semispinalis capitis muscle

Fourth cervical vertebra

Longus colli muscle

Middle pharyngeal constrictor muscle

Thyroid cartilage

FIGURE 7.31   The sagittal sections of the (A) nasal cavity and (B) maxillary sinus.

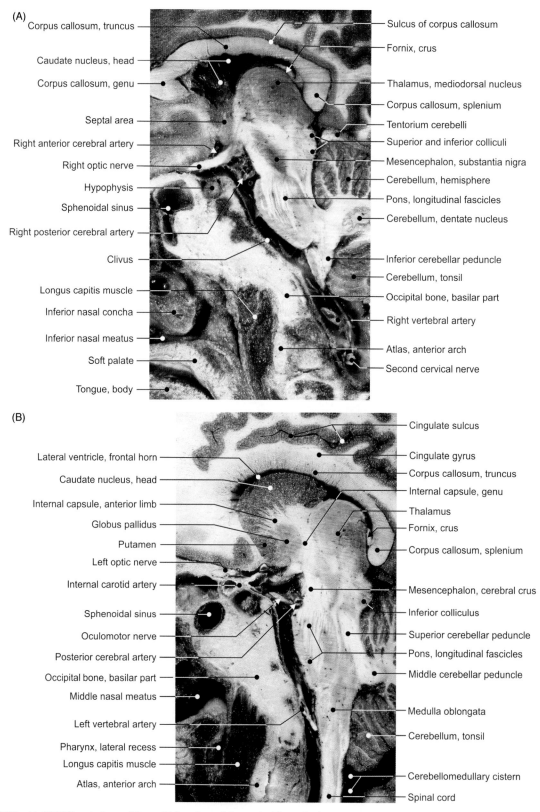

(A)
Corpus callosum, truncus
Caudate nucleus, head
Corpus callosum, genu
Septal area
Right anterior cerebral artery
Right optic nerve
Hypophysis
Sphenoidal sinus
Right posterior cerebral artery
Clivus
Longus capitis muscle
Inferior nasal concha
Inferior nasal meatus
Soft palate
Tongue, body

Sulcus of corpus callosum
Fornix, crus
Thalamus, mediodorsal nucleus
Corpus callosum, splenium
Tentorium cerebelli
Superior and inferior colliculi
Mesencephalon, substantia nigra
Cerebellum, hemisphere
Pons, longitudinal fascicles
Cerebellum, dentate nucleus
Inferior cerebellar peduncle
Cerebellum, tonsil
Occipital bone, basilar part
Right vertebral artery
Atlas, anterior arch
Second cervical nerve

(B)
Lateral ventricle, frontal horn
Caudate nucleus, head
Internal capsule, anterior limb
Globus pallidus
Putamen
Left optic nerve
Internal carotid artery
Sphenoidal sinus
Oculomotor nerve
Posterior cerebral artery
Occipital bone, basilar part
Middle nasal meatus
Left vertebral artery
Pharynx, lateral recess
Longus capitis muscle
Atlas, anterior arch

Cingulate sulcus
Cingulate gyrus
Corpus callosum, truncus
Internal capsule, genu
Thalamus
Fornix, crus
Corpus callosum, splenium
Mesencephalon, cerebral crus
Inferior colliculus
Superior cerebellar peduncle
Pons, longitudinal fascicles
Middle cerebellar peduncle
Medulla oblongata
Cerebellum, tonsil
Cerebellomedullary cistern
Spinal cord

FIGURE 7.32 (A–B) Different views of the sagittal sections of the cerebral hypophysis.

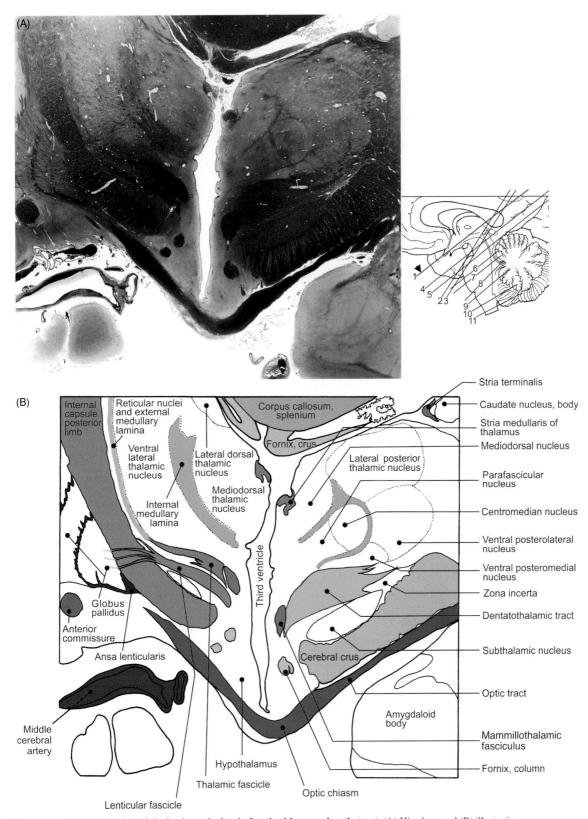

(A)

(B)

Internal
capsule,
posterior
limb

Reticular nuclei
and external
medullary
lamina

Corpus callosum,
splenium

Fornix, crus

Ventral
lateral
thalamic
nucleus

Lateral dorsal
thalamic
nucleus

Lateral posterior
thalamic nucleus

Mediodorsal
thalamic
nucleus

Internal
medullary
lamina

Third ventricle

Globus
pallidus

Anterior
commissure

Ansa lenticularis

Cerebral crus

Middle
cerebral
artery

Amygdaloid
body

Hypothalamus

Thalamic fascicle

Optic chiasm

Lenticular fascicle

Stria terminalis

Caudate nucleus, body

Stria medullaris of
thalamus

Mediodorsal nucleus

Parafascicular
nucleus

Centromedian nucleus

Ventral posterolateral
nucleus

Ventral posteromedial
nucleus

Zona incerta

Dentatothalamic tract

Subthalamic nucleus

Optic tract

Mammillothalamic
fasciculus

Fornix, column

FIGURE 7.33 **Transverse section of the brain at the level of optic chiasm and optic tract.** (A) Histology and (B) illustration.

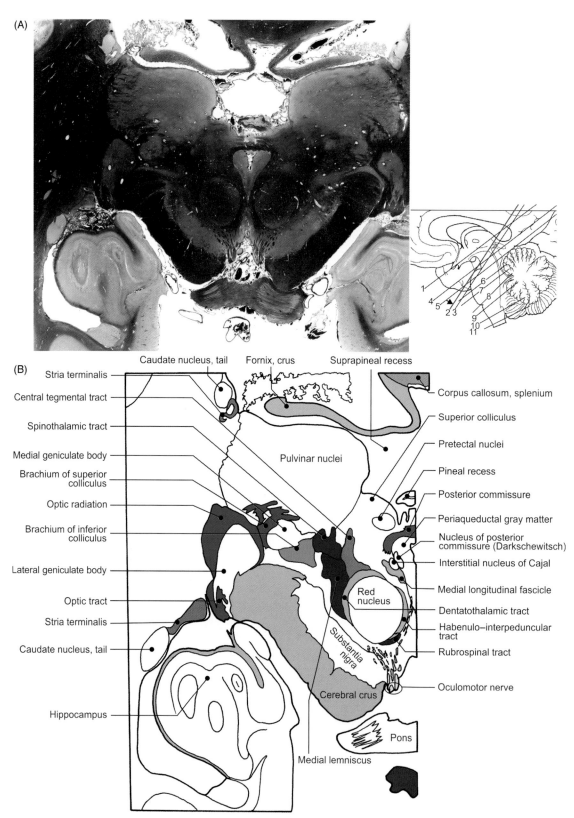

FIGURE 7.34 **Transverse section of the brain at the level of red nucleus.** (A) Histology and (B) illustration.

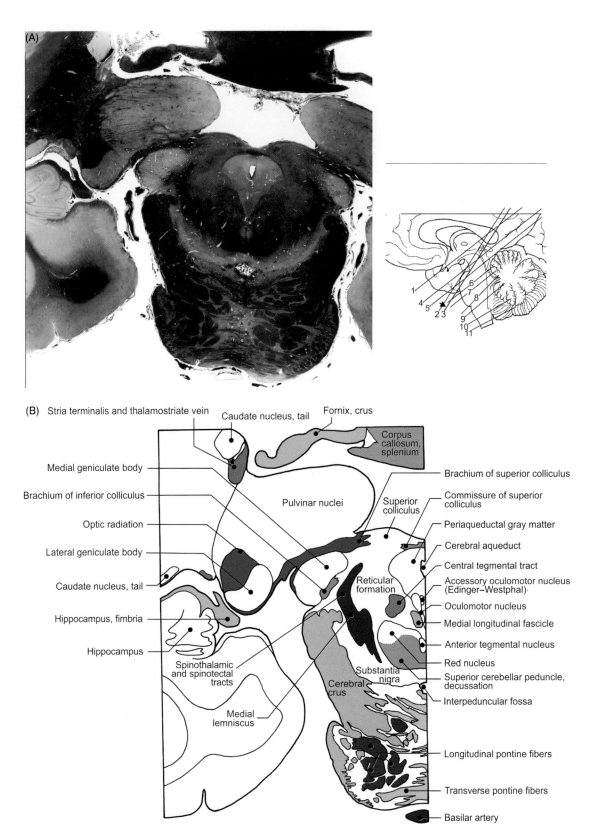

FIGURE 7.35   Transverse section of the brain at the level of lateral and medial geniculate bodies. (A) Histology and (B) illustration.

(A)

(B)

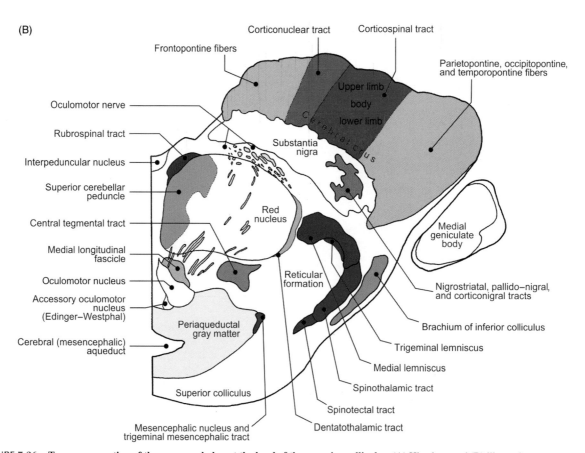

FIGURE 7.36 **Transverse section of the mesencephalon at the level of the superior colliculus.** (A) Histology and (B) illustration.

(A)

(B)

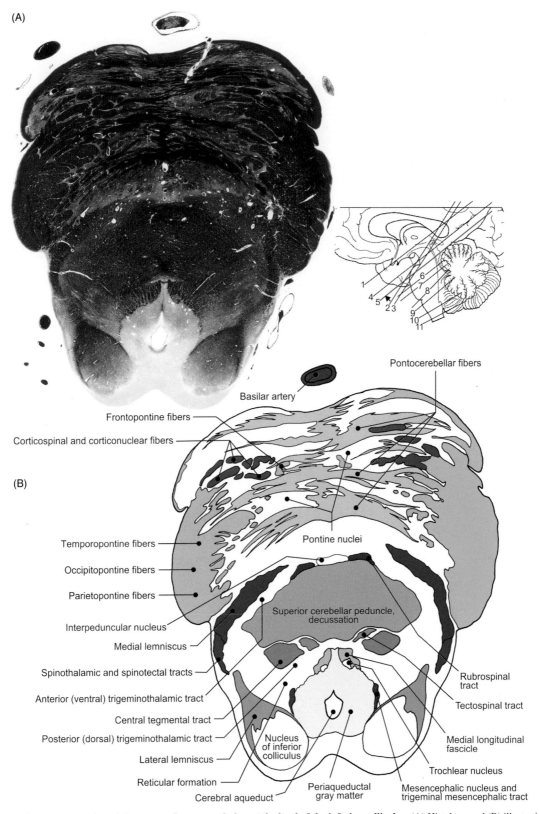

Basilar artery

Pontocerebellar fibers

Frontopontine fibers

Corticospinal and corticonuclear fibers

Temporopontine fibers

Occipitopontine fibers

Parietopontine fibers

Interpeduncular nucleus

Medial lemniscus

Spinothalamic and spinotectal tracts

Anterior (ventral) trigeminothalamic tract

Central tegmental tract

Posterior (dorsal) trigeminothalamic tract

Lateral lemniscus

Reticular formation

Cerebral aqueduct

Pontine nuclei

Superior cerebellar peduncle, decussation

Nucleus of inferior colliculus

Periaqueductal gray matter

Rubrospinal tract

Tectospinal tract

Medial longitudinal fascicle

Trochlear nucleus

Mesencephalic nucleus and trigeminal mesencephalic tract

FIGURE 7.37 **Transverse section of the pons and mesencephalon at the level of the inferior colliculus.** (A) Histology and (B) illustration.

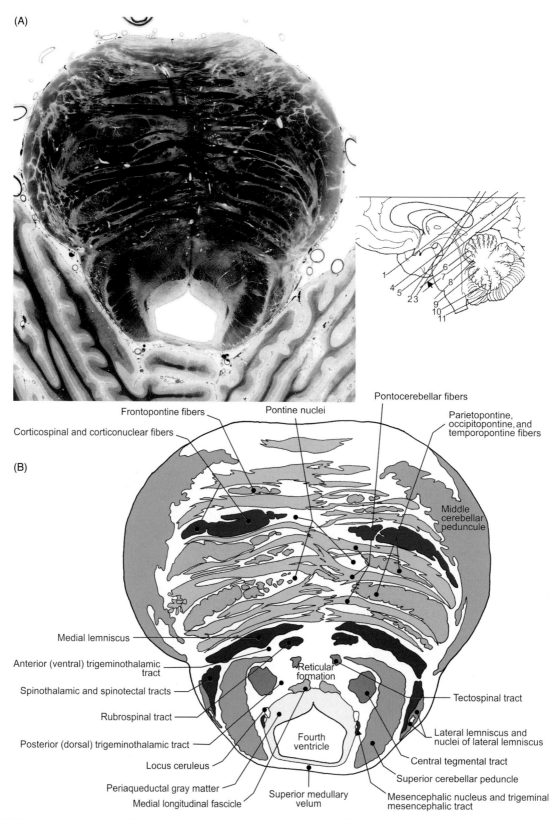

(A)

(B)

Frontopontine fibers

Corticospinal and corticonuclear fibers

Pontine nuclei

Pontocerebellar fibers

Parietopontine, occipitopontine, and temporopontine fibers

Middle cerebellar peduncule

Medial lemniscus

Anterior (ventral) trigeminothalamic tract

Spinothalamic and spinotectal tracts

Rubrospinal tract

Posterior (dorsal) trigeminothalamic tract

Locus ceruleus

Periaqueductal gray matter

Medial longitudinal fascicle

Reticular formation

Fourth ventricle

Superior medullary velum

Tectospinal tract

Lateral lemniscus and nuclei of lateral lemniscus

Central tegmental tract

Superior cerebellar peduncle

Mesencephalic nucleus and trigeminal mesencephalic tract

FIGURE 7.38   Transverse section of the upper segment of pons. (A) Histology and (B) illustration.

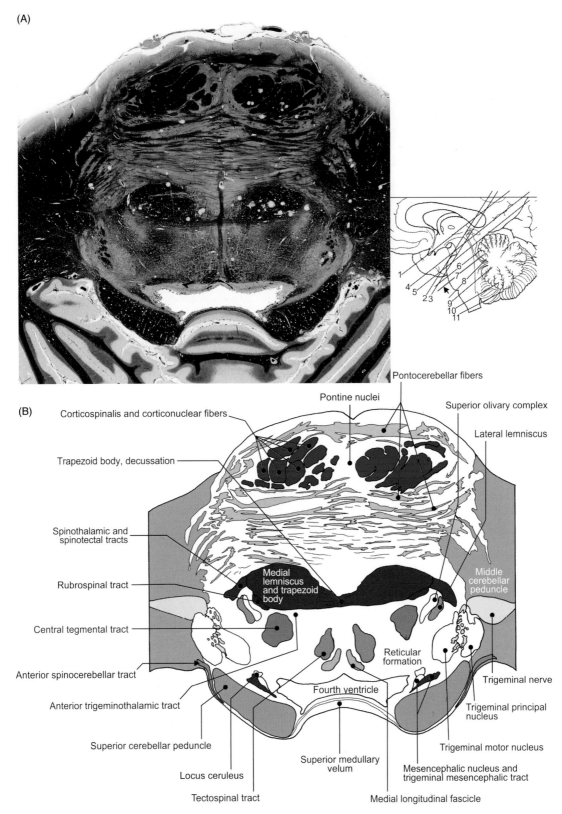

FIGURE 7.39   Transverse section of the middle third of pons. (A) Histology and (B) illustration.

(A)

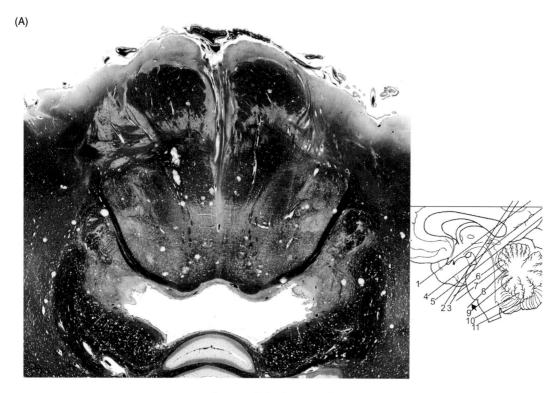

(B)

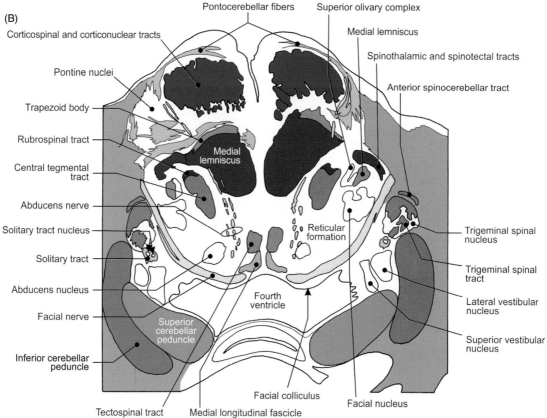

Pontocerebellar fibers

Superior olivary complex

Corticospinal and corticonuclear tracts

Medial lemniscus

Spinothalamic and spinotectal tracts

Pontine nuclei

Anterior spinocerebellar tract

Trapezoid body

Rubrospinal tract

Medial lemniscus

Central tegmental tract

Abducens nerve

Solitary tract nucleus

Reticular formation

Trigeminal spinal nucleus

Solitary tract

Trigeminal spinal tract

Abducens nucleus

Lateral vestibular nucleus

Facial nerve

Fourth ventricle

Superior cerebellar peduncle

Superior vestibular nucleus

Inferior cerebellar peduncle

Facial colliculus

Facial nucleus

Tectospinal tract

Medial longitudinal fascicle

FIGURE 7.40 **Transverse section of pons at the level of the facial nerve nucleus.** (A) Histology and (B) illustration.

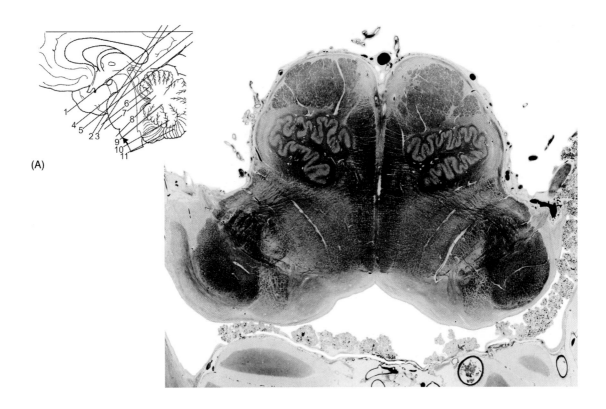

(A)

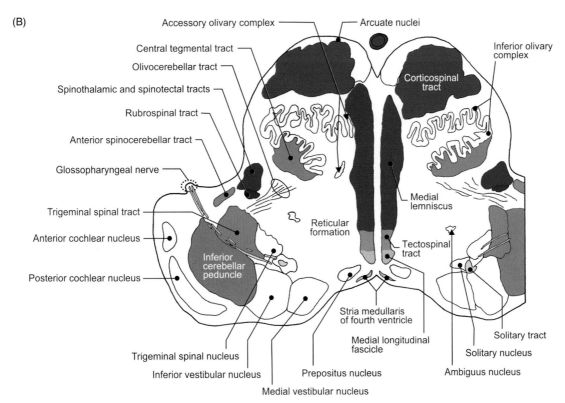

(B)

FIGURE 7.41   **Transverse section of the uppermost segment of medulla oblongata.** (A) Histology and (B) illustration.

(A)

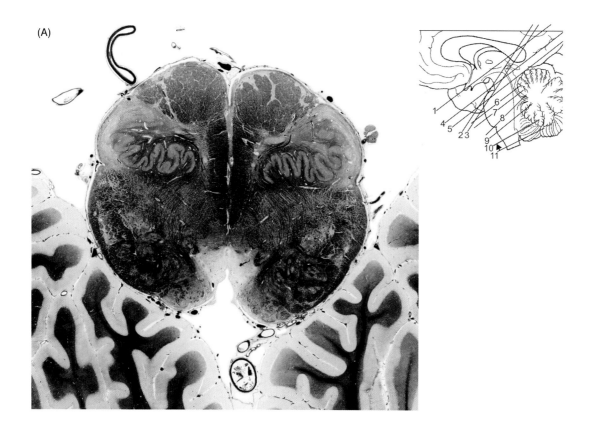

(B)

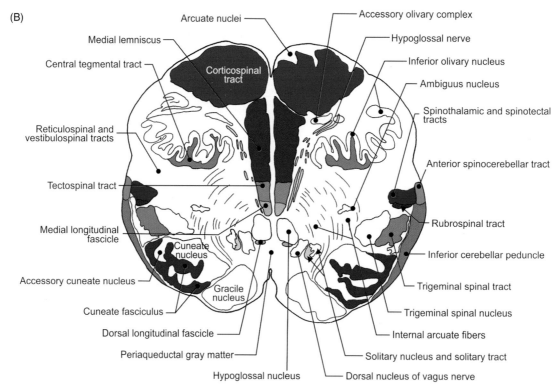

FIGURE 7.42 Transverse section through the middle segment of medulla oblongata. (A) Histology and (B) illustration.

(A)

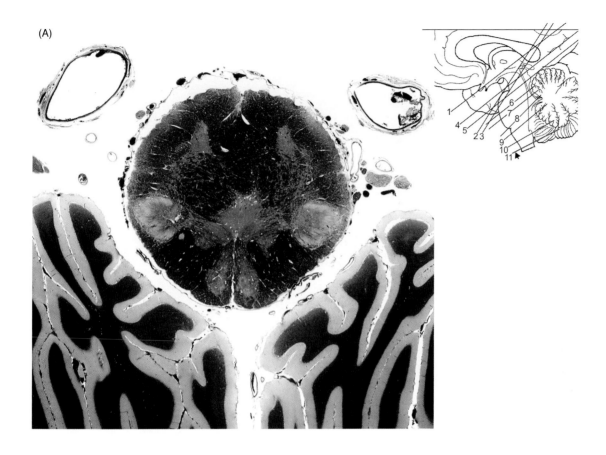

(B)

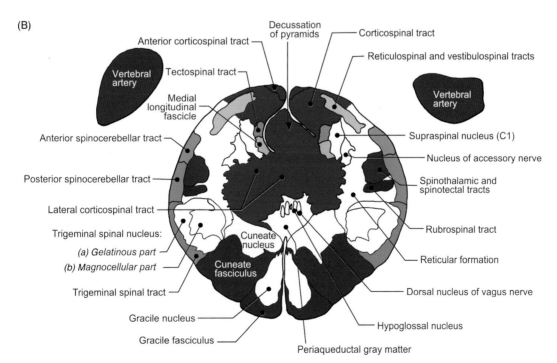

FIGURE 7.43   **Transverse section of medulla oblongata through the decussation of pyramids.** (A) Histology and (B) illustration.

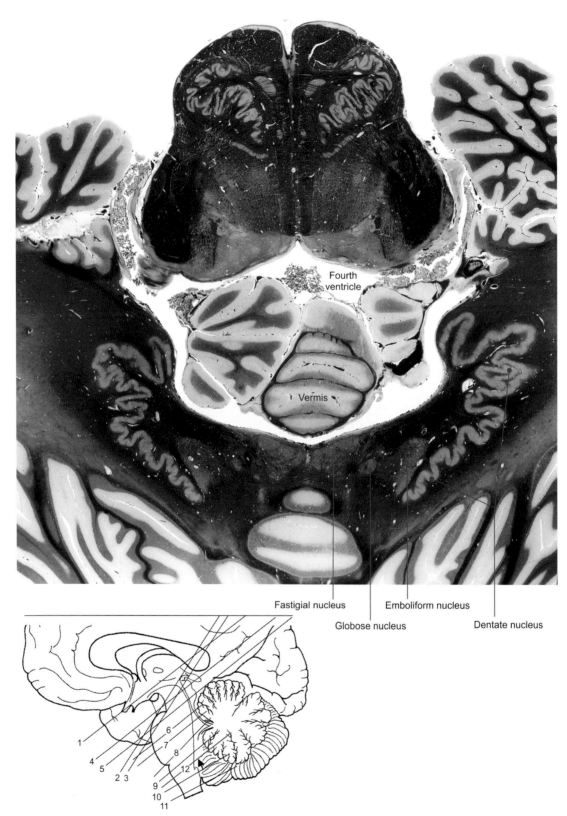

Fourth
ventricle

Vermis

Fastigial nucleus   Emboliform nucleus

Globose nucleus   Dentate nucleus

FIGURE 7.44 **Transverse section of cerebellar nuclei.**

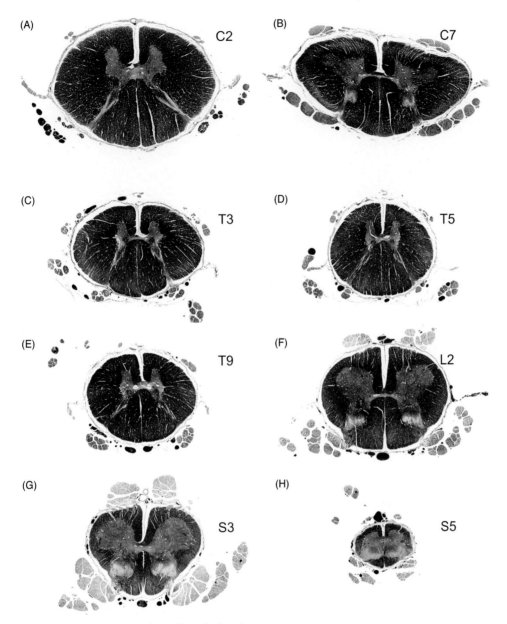

FIGURE 7.45   (A–H) Sequential transverse sections of the spinal cord.

(A)

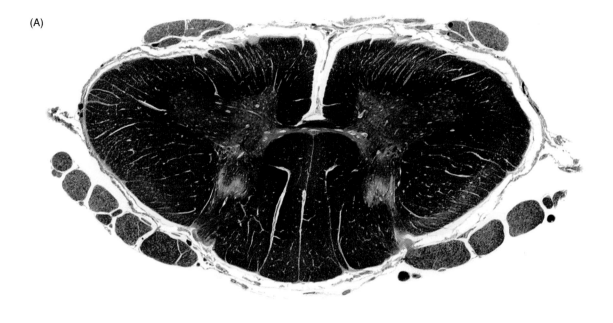

(B)

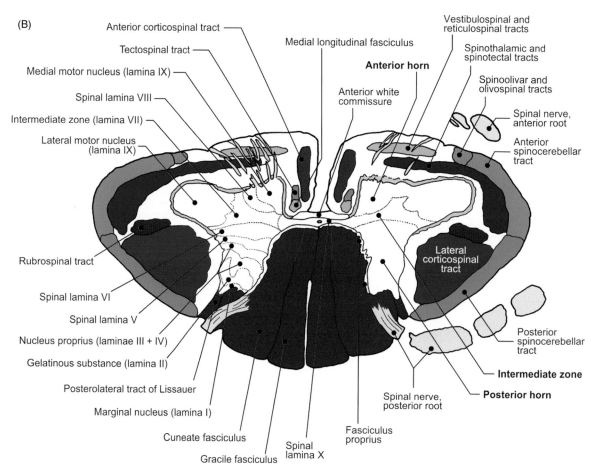

FIGURE 7.46 **Transverse section of C7 segment of the spinal cord.** (A) Histology and (B) illustration.

(A)

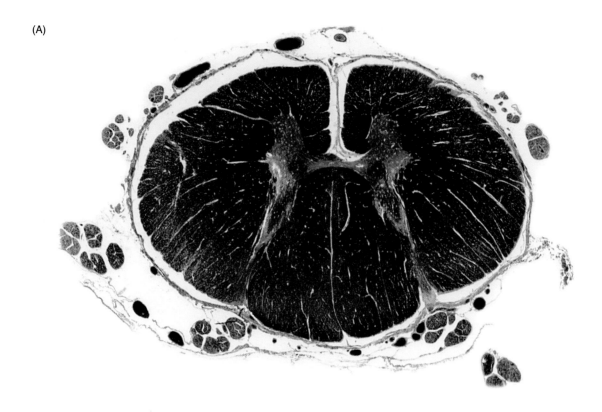

(B)

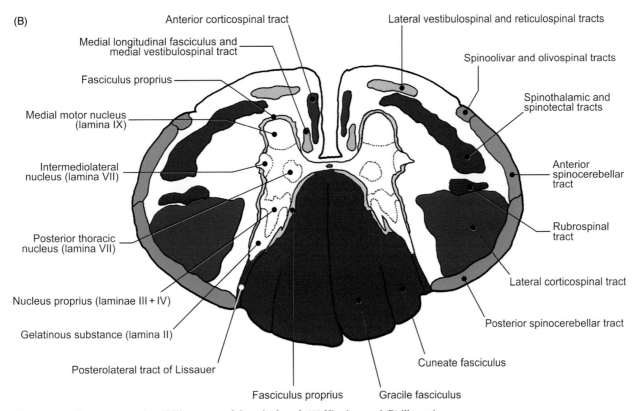

Anterior corticospinal tract

Lateral vestibulospinal and reticulospinal tracts

Medial longitudinal fasciculus and medial vestibulospinal tract

Spinoolivar and olivospinal tracts

Fasciculus proprius

Spinothalamic and spinotectal tracts

Medial motor nucleus (lamina IX)

Intermediolateral nucleus (lamina VII)

Anterior spinocerebellar tract

Posterior thoracic nucleus (lamina VII)

Rubrospinal tract

Nucleus proprius (laminae III + IV)

Lateral corticospinal tract

Gelatinous substance (lamina II)

Posterior spinocerebellar tract

Posterolateral tract of Lissauer

Fasciculus proprius

Gracile fasciculus

Cuneate fasciculus

FIGURE 7.47 **Transverse section of T3 segment of the spinal cord.** (A) Histology and (B) illustration.

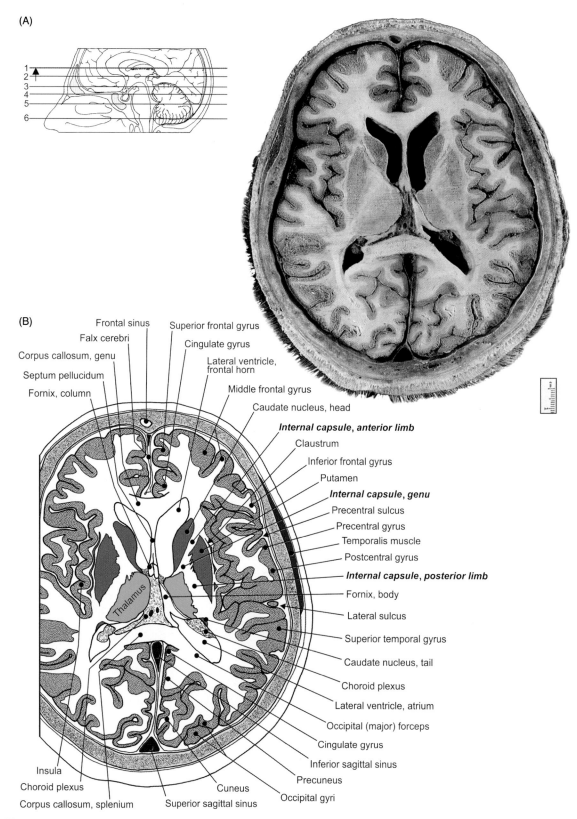

(A)

(B)

Frontal sinus
Falx cerebri
Corpus callosum, genu
Septum pellucidum
Fornix, column
Superior frontal gyrus
Cingulate gyrus
Lateral ventricle, frontal horn
Middle frontal gyrus
Caudate nucleus, head
Internal capsule, anterior limb
Claustrum
Inferior frontal gyrus
Putamen
Internal capsule, genu
Precentral sulcus
Precentral gyrus
Temporalis muscle
Postcentral gyrus
Internal capsule, posterior limb
Fornix, body
Lateral sulcus
Superior temporal gyrus
Caudate nucleus, tail
Choroid plexus
Lateral ventricle, atrium
Occipital (major) forceps
Cingulate gyrus
Inferior sagittal sinus
Precuneus
Occipital gyri
Thalamus
Insula
Choroid plexus
Corpus callosum, splenium
Cuneus
Superior sagittal sinus

FIGURE 7.48   Transverse section of the genu and splenium of corpus callosum. (A) Section and (B) illustration.

(A)

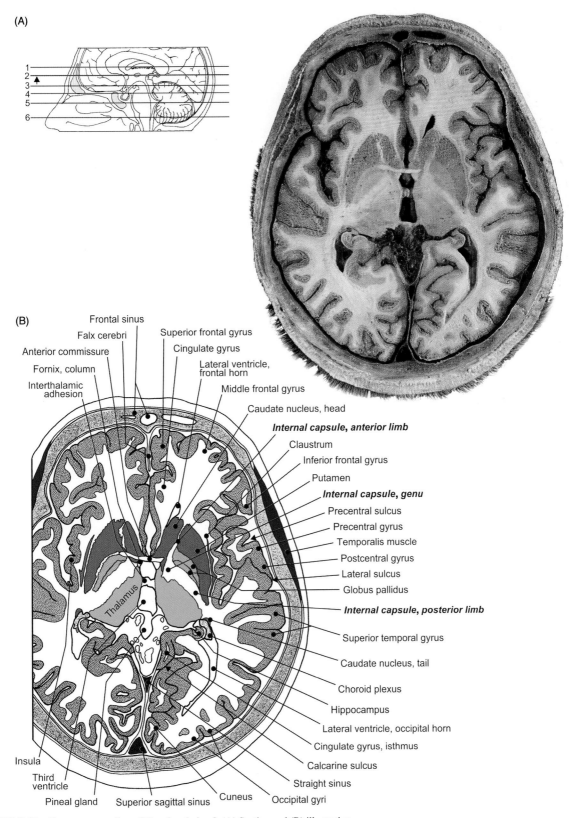

(B)

Frontal sinus
Falx cerebri
Anterior commissure
Fornix, column
Interthalamic adhesion
Superior frontal gyrus
Cingulate gyrus
Lateral ventricle, frontal horn
Middle frontal gyrus
Caudate nucleus, head
Internal capsule, anterior limb
Claustrum
Inferior frontal gyrus
Putamen
Internal capsule, genu
Precentral sulcus
Precentral gyrus
Temporalis muscle
Postcentral gyrus
Lateral sulcus
Globus pallidus
Internal capsule, posterior limb
Superior temporal gyrus
Caudate nucleus, tail
Choroid plexus
Hippocampus
Lateral ventricle, occipital horn
Cingulate gyrus, isthmus
Calcarine sulcus
Straight sinus
Occipital gyri
Cuneus
Superior sagittal sinus
Pineal gland
Third ventricle
Insula
Thalamus

FIGURE 7.49   **Transverse section of the pineal gland.** (A) Section and (B) illustration.

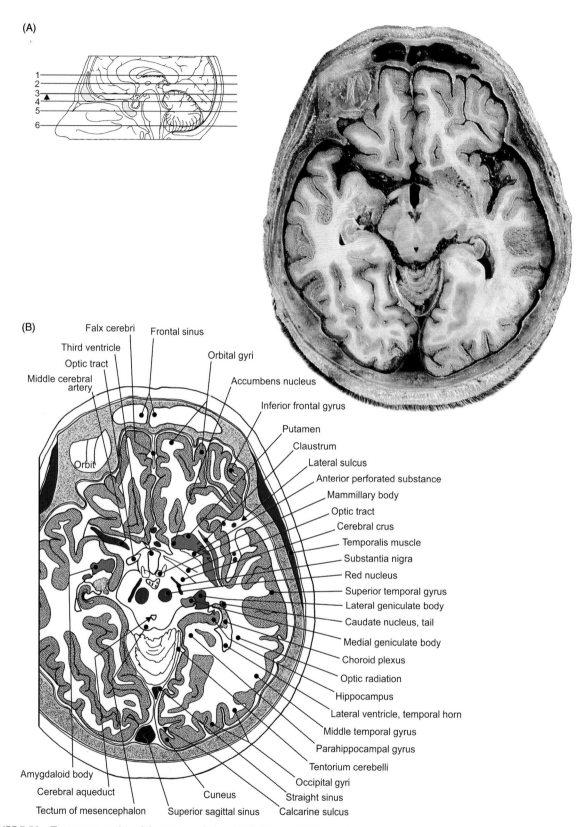

(A)

(B)

Falx cerebri
Frontal sinus
Third ventricle
Optic tract
Orbital gyri
Middle cerebral artery
Accumbens nucleus
Inferior frontal gyrus
Orbit
Putamen
Claustrum
Lateral sulcus
Anterior perforated substance
Mammillary body
Optic tract
Cerebral crus
Temporalis muscle
Substantia nigra
Red nucleus
Superior temporal gyrus
Lateral geniculate body
Caudate nucleus, tail
Medial geniculate body
Choroid plexus
Optic radiation
Hippocampus
Lateral ventricle, temporal horn
Middle temporal gyrus
Parahippocampal gyrus
Tentorium cerebelli
Amygdaloid body
Occipital gyri
Cerebral aqueduct
Cuneus
Straight sinus
Tectum of mesencephalon
Superior sagittal sinus
Calcarine sulcus

FIGURE 7.50   Transverse section of the mesencephalon. (A) Section and (B) illustration.

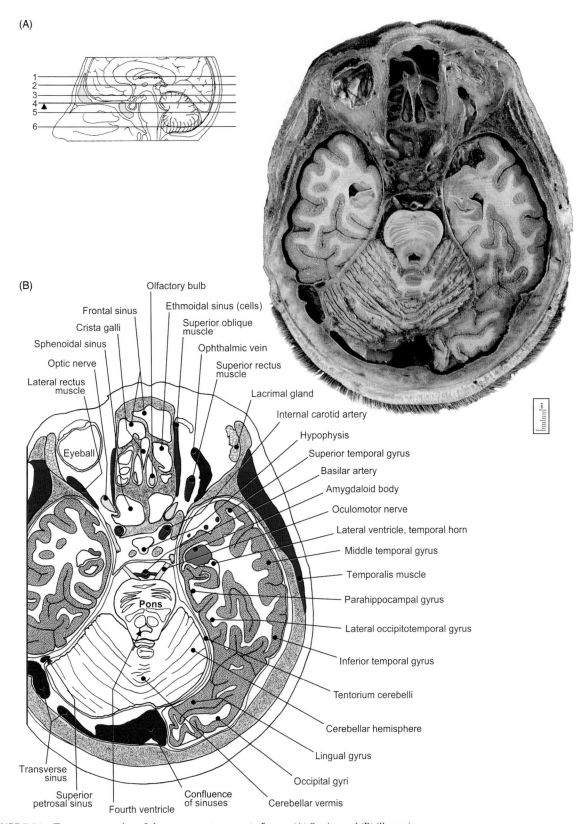

(A)

(B)

Olfactory bulb

Frontal sinus

Crista galli

Sphenoidal sinus

Optic nerve

Lateral rectus muscle

Ethmoidal sinus (cells)

Superior oblique muscle

Ophthalmic vein

Superior rectus muscle

Lacrimal gland

Internal carotid artery

Hypophysis

Superior temporal gyrus

Basilar artery

Amygdaloid body

Oculomotor nerve

Lateral ventricle, temporal horn

Middle temporal gyrus

Temporalis muscle

Parahippocampal gyrus

Lateral occipitotemporal gyrus

Inferior temporal gyrus

Tentorium cerebelli

Cerebellar hemisphere

Lingual gyrus

Occipital gyri

Cerebellar vermis

Eyeball

Pons

Transverse sinus

Superior petrosal sinus

Fourth ventricle

Confluence of sinuses

FIGURE 7.51   Transverse section of the uppermost segment of pons. (A) Section and (B) illustration.

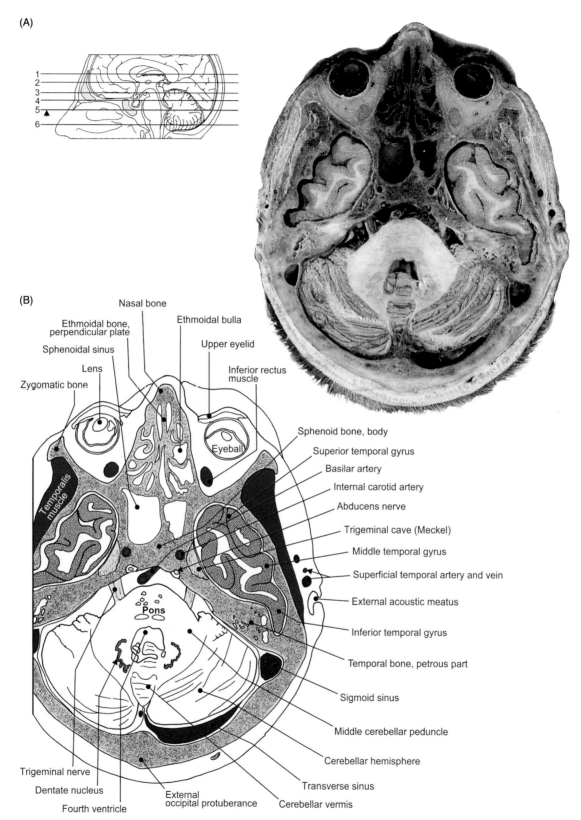

(A)

(B)

Nasal bone

Ethmoidal bone,
perpendicular plate

Ethmoidal bulla

Sphenoidal sinus

Upper eyelid

Lens

Inferior rectus
muscle

Zygomatic bone

Eyeball

Temporalis muscle

Sphenoid bone, body

Superior temporal gyrus

Basilar artery

Internal carotid artery

Abducens nerve

Trigeminal cave (Meckel)

Middle temporal gyrus

Superficial temporal artery and vein

External acoustic meatus

Inferior temporal gyrus

Pons

Temporal bone, petrous part

Sigmoid sinus

Middle cerebellar peduncle

Cerebellar hemisphere

Transverse sinus

Trigeminal nerve

Dentate nucleus

External
occipital protuberance

Cerebellar vermis

Fourth ventricle

FIGURE 7.52   Transverse section of the lowermost segment of pons. (A) Section and (B) illustration.

(A)

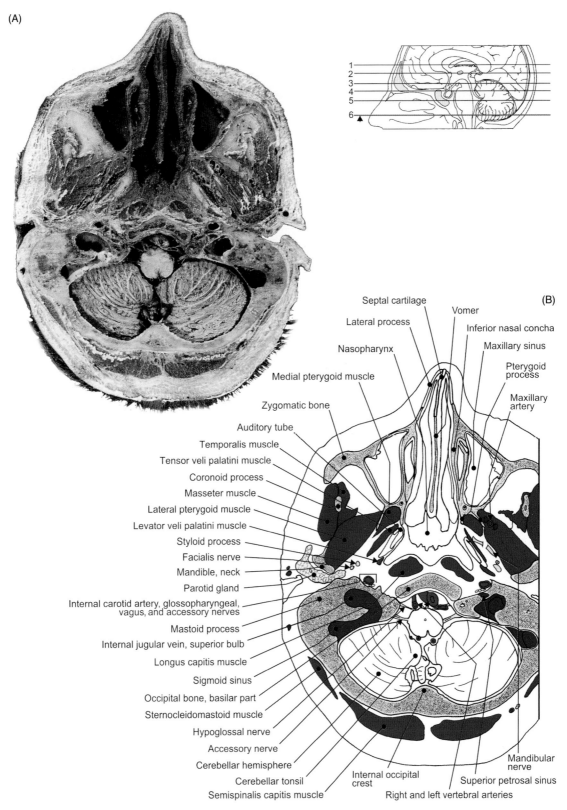

Septal cartilage

Lateral process

Nasopharynx

Medial pterygoid muscle

Zygomatic bone

Auditory tube

Temporalis muscle

Tensor veli palatini muscle

Coronoid process

Masseter muscle

Lateral pterygoid muscle

Levator veli palatini muscle

Styloid process

Facialis nerve

Mandible, neck

Parotid gland

Internal carotid artery, glossopharyngeal, vagus, and accessory nerves

Mastoid process

Internal jugular vein, superior bulb

Longus capitis muscle

Sigmoid sinus

Occipital bone, basilar part

Sternocleidomastoid muscle

Hypoglossal nerve

Accessory nerve

Cerebellar hemisphere

Cerebellar tonsil

Semispinalis capitis muscle

Vomer

Inferior nasal concha

Maxillary sinus

Pterygoid process

Maxillary artery

(B)

Mandibular nerve

Internal occipital crest

Superior petrosal sinus

Right and left vertebral arteries

FIGURE 7.53 **Transverse section of medulla oblongata.** (A) Section and (B) illustration.

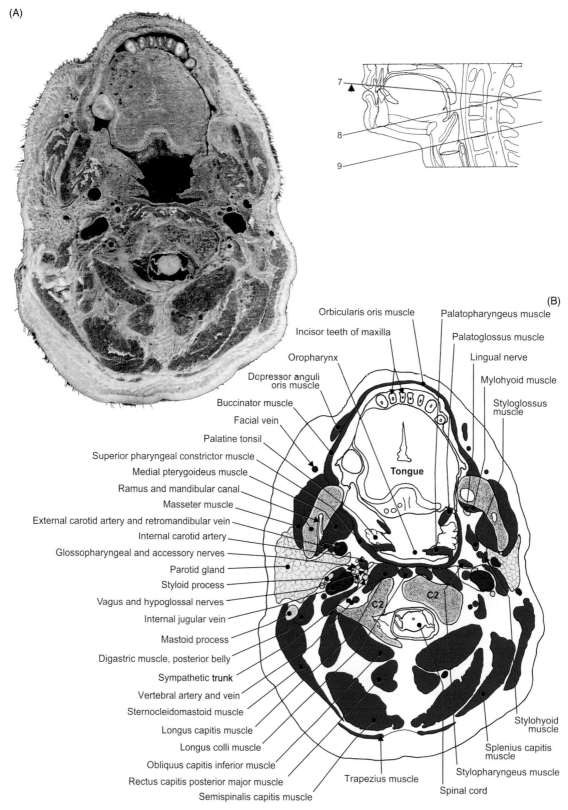

(A)

(B)

Orbicularis oris muscle

Incisor teeth of maxilla

Oropharynx

Depressor anguli
oris muscle

Buccinator muscle

Facial vein

Palatine tonsil

Superior pharyngeal constrictor muscle

Medial pterygoideus muscle

Ramus and mandibular canal

Masseter muscle

External carotid artery and retromandibular vein

Internal carotid artery

Glossopharyngeal and accessory nerves

Parotid gland

Styloid process

Vagus and hypoglossal nerves

Internal jugular vein

Mastoid process

Digastric muscle, posterior belly

Sympathetic trunk

Vertebral artery and vein

Sternocleidomastoid muscle

Longus capitis muscle

Longus colli muscle

Obliquus capitis inferior muscle

Rectus capitis posterior major muscle

Semispinalis capitis muscle

Palatopharyngeus muscle

Palatoglossus muscle

Lingual nerve

Mylohyoid muscle

Styloglossus
muscle

Tongue

Stylohyoid
muscle

Splenius capitis
muscle

Stylopharyngeus muscle

Spinal cord

Trapezius muscle

C2

FIGURE 7.54 **Transverse section of the oral cavity at the level of C2 vertebra.** (A) Section and (B) illustration.

(A)

(B)

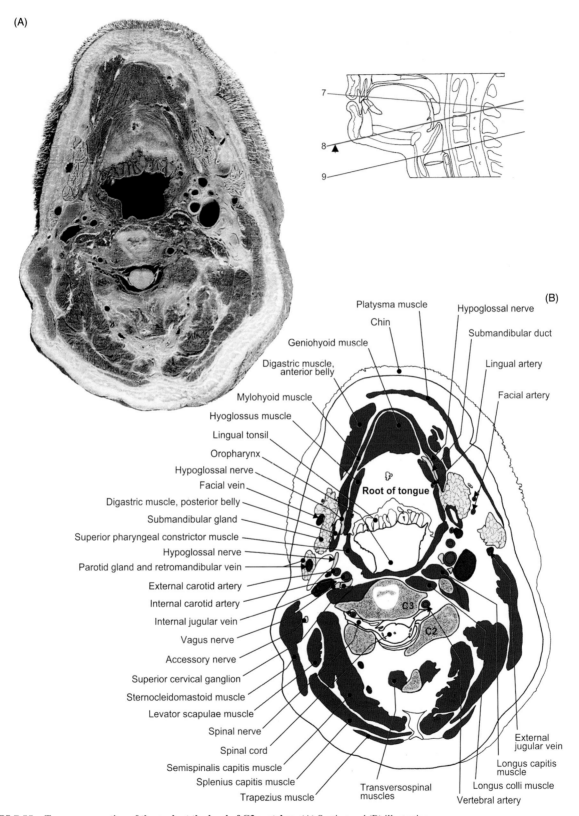

Platysma muscle
Chin
Geniohyoid muscle
Digastric muscle, anterior belly
Mylohyoid muscle
Hyoglossus muscle
Lingual tonsil
Oropharynx
Hypoglossal nerve
Facial vein
Digastric muscle, posterior belly
Submandibular gland
Superior pharyngeal constrictor muscle
Hypoglossal nerve
Parotid gland and retromandibular vein
External carotid artery
Internal carotid artery
Internal jugular vein
Vagus nerve
Accessory nerve
Superior cervical ganglion
Sternocleidomastoid muscle
Levator scapulae muscle
Spinal nerve
Spinal cord
Semispinalis capitis muscle
Splenius capitis muscle
Trapezius muscle

Hypoglossal nerve
Submandibular duct
Lingual artery
Facial artery

Root of tongue

C3
C2

External jugular vein
Longus capitis muscle
Longus colli muscle
Vertebral artery

Transversospinal muscles

FIGURE 7.55 **Transverse section of the neck at the level of C3 vertebra.** (A) Section and (B) illustration.

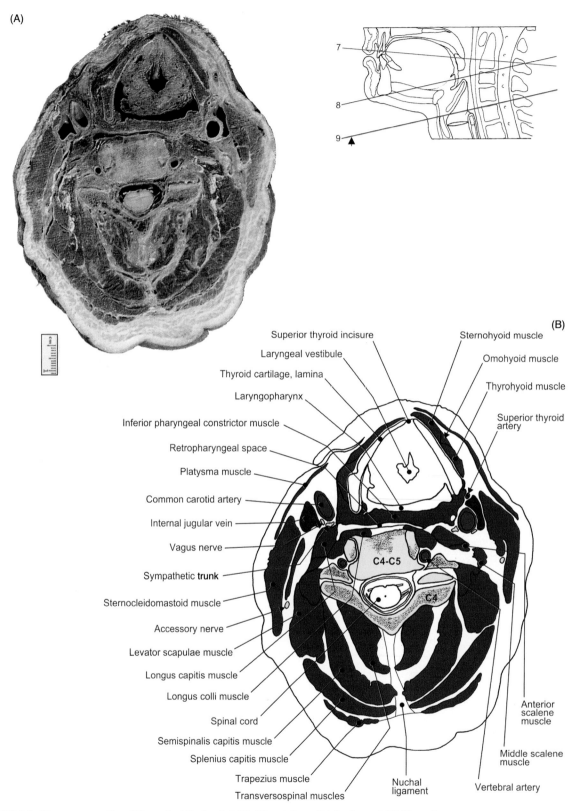

(A)

(B)

Superior thyroid incisure

Laryngeal vestibule

Thyroid cartilage, lamina

Laryngopharynx

Inferior pharyngeal constrictor muscle

Retropharyngeal space

Platysma muscle

Common carotid artery

Internal jugular vein

Vagus nerve

Sympathetic **trunk**

Sternocleidomastoid muscle

Accessory nerve

Levator scapulae muscle

Longus capitis muscle

Longus colli muscle

Spinal cord

Semispinalis capitis muscle

Splenius capitis muscle

Trapezius muscle

Transversospinal muscles

Sternohyoid muscle

Omohyoid muscle

Thyrohyoid muscle

Superior thyroid artery

C4-C5

C4

Nuchal ligament

Anterior scalene muscle

Middle scalene muscle

Vertebral artery

FIGURE 7.56 Transverse section of the neck at the level of C4–C5 vertebrae. (A) Section and (B) illustration.

Chapter 8

Vascularization of Head and Neck and the Cranial Central Nervous System

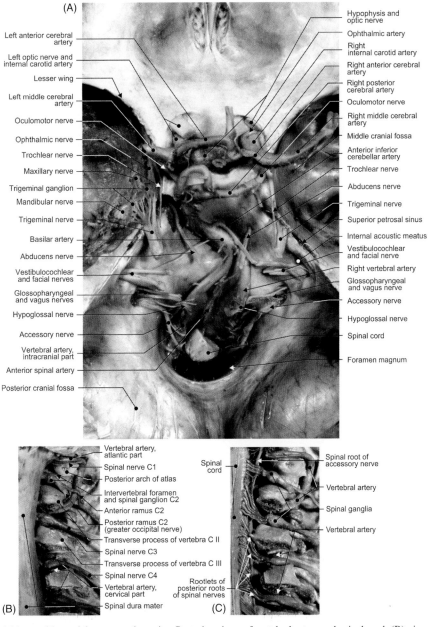

(A)

Left anterior cerebral artery
Left optic nerve and internal carotid artery
Lesser wing
Left middle cerebral artery
Oculomotor nerve
Ophthalmic nerve
Trochlear nerve
Maxillary nerve
Trigeminal ganglion
Mandibular nerve
Trigeminal nerve
Basilar artery
Abducens nerve
Vestibulocochlear and facial nerves
Glossopharyngeal and vagus nerves
Hypoglossal nerve
Accessory nerve
Vertebral artery, intracranial part
Anterior spinal artery
Posterior cranial fossa

Hypophysis and optic nerve
Ophthalmic artery
Right internal carotid artery
Right anterior cerebral artery
Right posterior cerebral artery
Oculomotor nerve
Right middle cerebral artery
Middle cranial fossa
Anterior inferior cerebellar artery
Trochlear nerve
Abducens nerve
Trigeminal nerve
Superior petrosal sinus
Internal acoustic meatus
Vestibulocochlear and facial nerve
Right vertebral artery
Glossopharyngeal and vagus nerve
Accessory nerve
Hypoglossal nerve
Spinal cord
Foramen magnum

Vertebral artery, atlantic part
Spinal nerve C1
Posterior arch of atlas
Intervertebral foramen and spinal ganglion C2
Anterior ramus C2
Posterior ramus C2 (greater occipital nerve)
Transverse process of vertebra C II
Spinal nerve C3
Transverse process of vertebra C III
Spinal nerve C4
Vertebral artery, cervical part
Spinal dura mater

(B)

Spinal cord

Spinal root of accessory nerve
Vertebral artery
Spinal ganglia
Vertebral artery

Rootlets of posterior roots of spinal nerves

(C)

FIGURE 8.1 (A) Cranial base with cranial nerves and arteries. Posterior views of vertebral artery and spinal cord: (B) view with vertebral segments removed and (C) view with dura mater removed.

Atlas of the Human Body
Copyright © 2017 Elsevier Inc. All rights reserved.

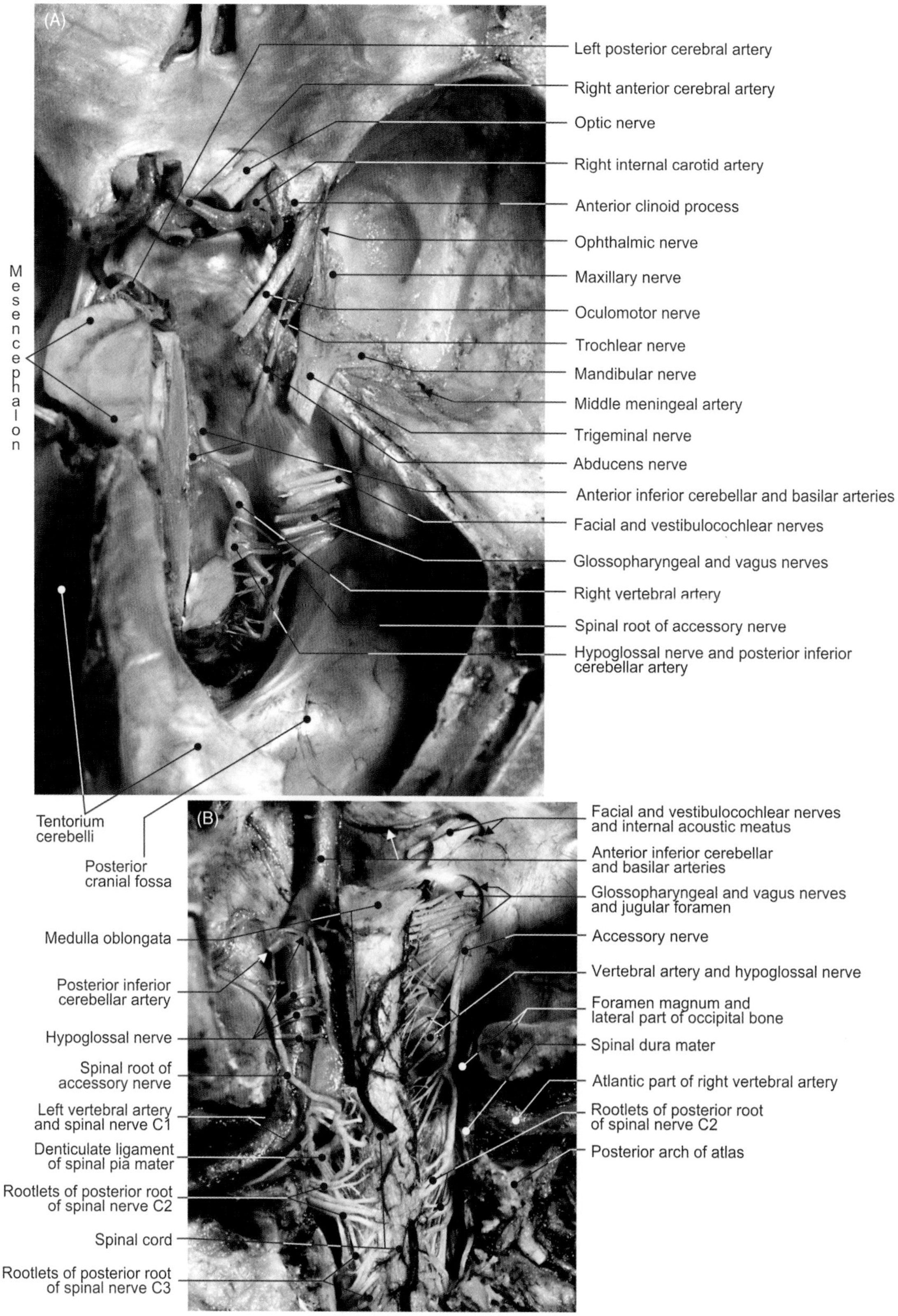

FIGURE 8.2 (A) Right half of the cranial base: cranial nerves and arteries. (B) Posterior view of the terminal segment of the vertebral artery: dissection.

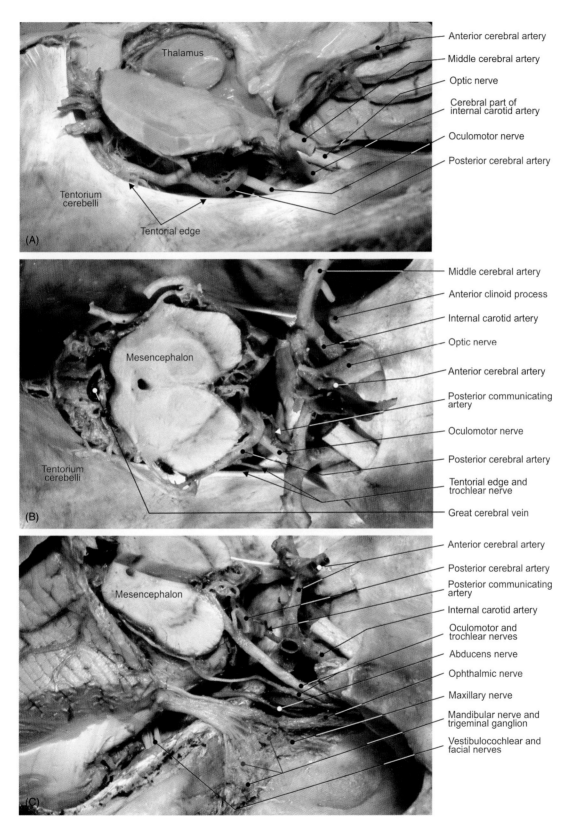

Anterior cerebral artery

Middle cerebral artery

Optic nerve

Cerebral part of
internal carotid artery

Oculomotor nerve

Posterior cerebral artery

Middle cerebral artery

Anterior clinoid process

Internal carotid artery

Optic nerve

Anterior cerebral artery

Posterior communicating
artery

Oculomotor nerve

Posterior cerebral artery

Tentorial edge and
trochlear nerve

Great cerebral vein

Anterior cerebral artery

Posterior cerebral artery

Posterior communicating
artery

Internal carotid artery

Oculomotor and
trochlear nerves

Abducens nerve

Ophthalmic nerve

Maxillary nerve

Mandibular nerve and
trigeminal ganglion

Vestibulocochlear and
facial nerves

Thalamus

Tentorium
cerebelli

Tentorial edge

(A)

Mesencephalon

Tentorium
cerebelli

(B)

Mesencephalon

(C)

FIGURE 8.3   (A) Right and (B) both hemispheres removed, view from the right side. (C) Relationships of structures on the cranial base.

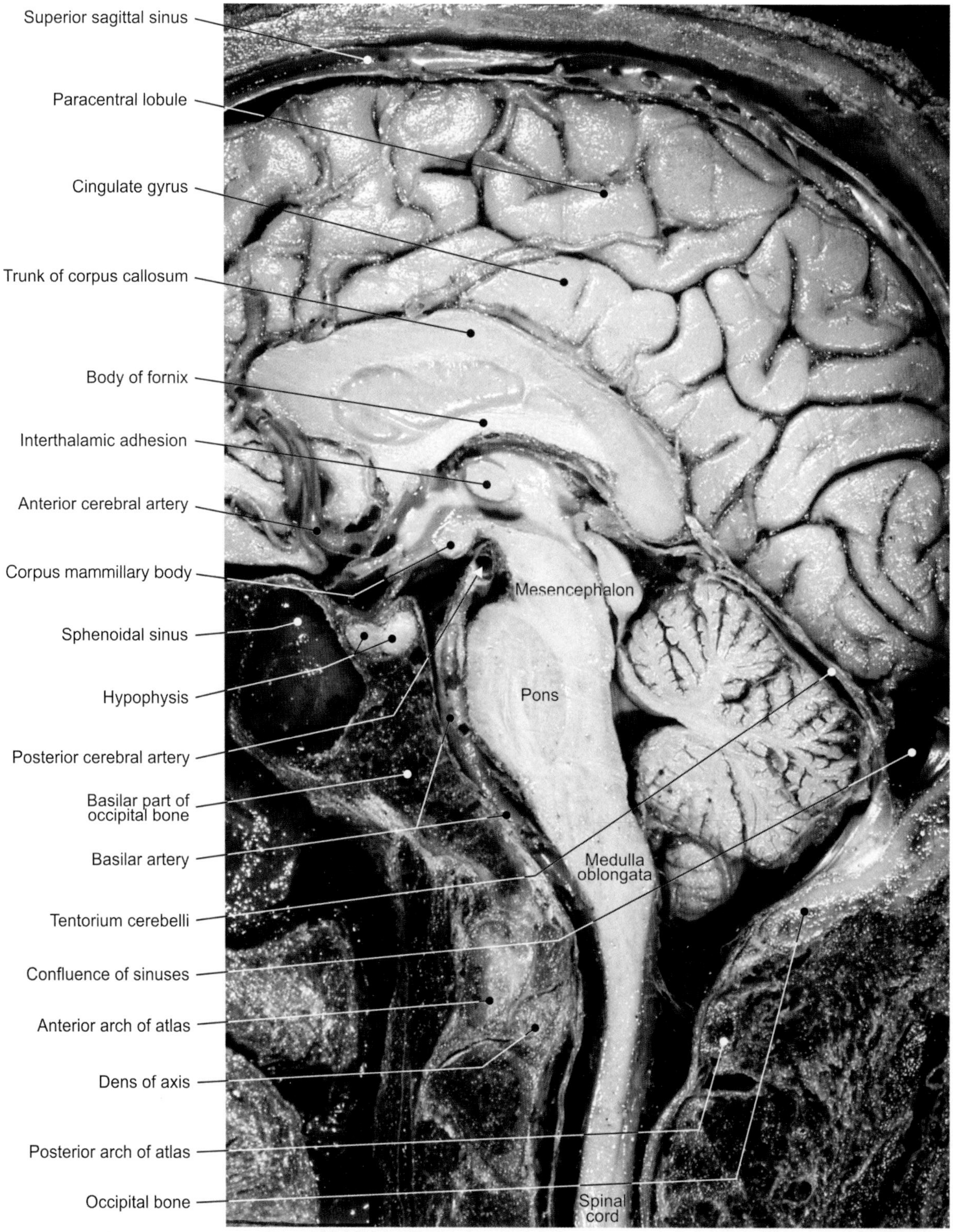

Superior sagittal sinus

Paracentral lobule

Cingulate gyrus

Trunk of corpus callosum

Body of fornix

Interthalamic adhesion

Anterior cerebral artery

Corpus mammillary body

Sphenoidal sinus

Hypophysis

Posterior cerebral artery

Basilar part of
occipital bone

Basilar artery

Tentorium cerebelli

Confluence of sinuses

Anterior arch of atlas

Dens of axis

Posterior arch of atlas

Occipital bone

Mesencephalon

Pons

Medulla
oblongata

Spinal
cord

FIGURE 8.4 **The midsagittal section of the head. Internal aspect of the right hemisphere.**

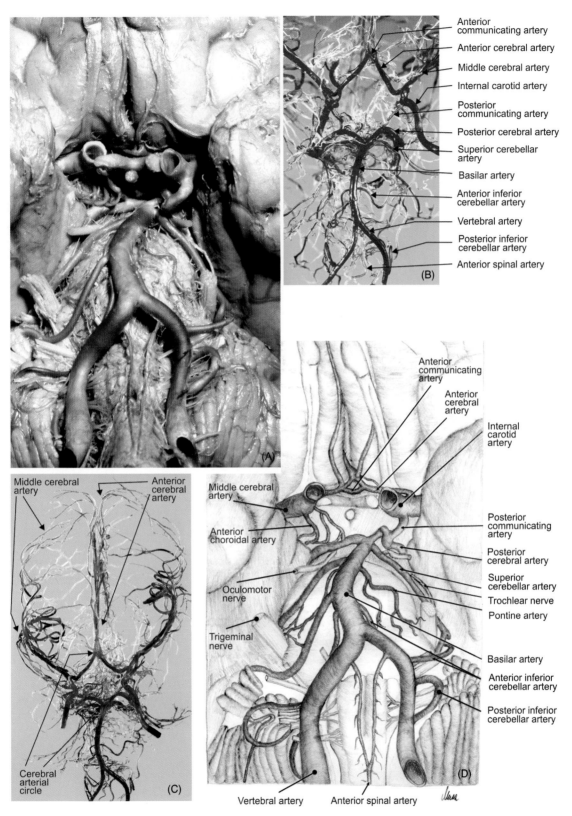

FIGURE 8.5 Arterial distribution on the base of the cranial central nervous system. (A) Formalin preparation, (B–C) corrosion casts, and (D) illustration.

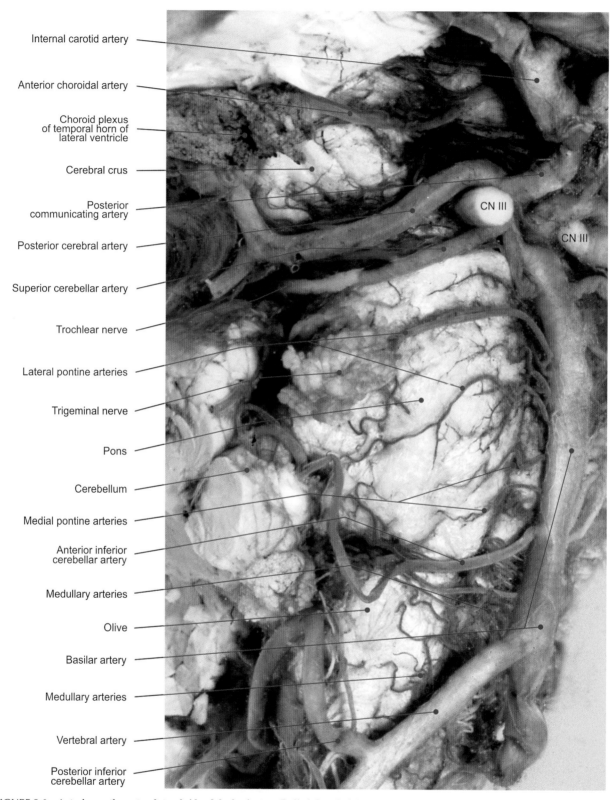

Internal carotid artery

Anterior choroidal artery

Choroid plexus
of temporal horn of
lateral ventricle

Cerebral crus

Posterior
communicating artery

Posterior cerebral artery

Superior cerebellar artery

Trochlear nerve

Lateral pontine arteries

Trigeminal nerve

Pons

Cerebellum

Medial pontine arteries

Anterior inferior
cerebellar artery

Medullary arteries

Olive

Basilar artery

Medullary arteries

Vertebral artery

Posterior inferior
cerebellar artery

CN III

CN III

FIGURE 8.6   **Arteries on the anterolateral side of the brainstem (India ink–gelatin).**

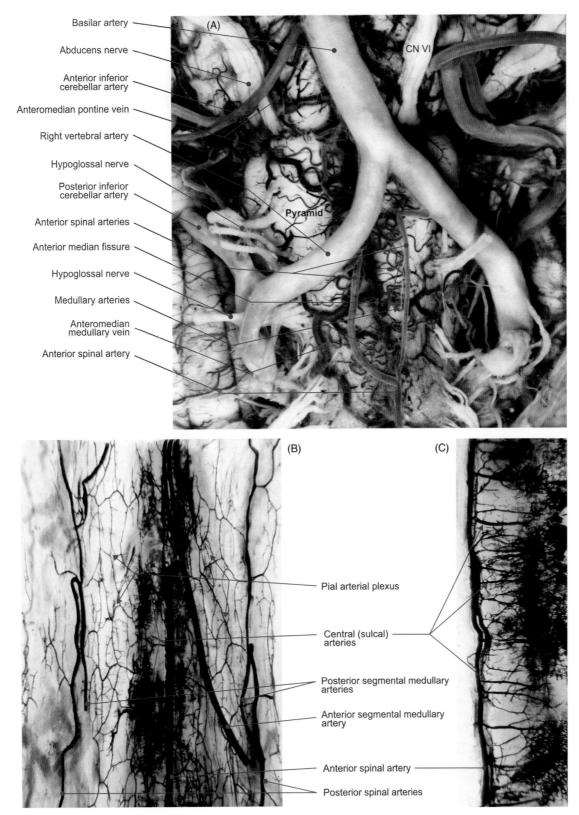

Basilar artery

Abducens nerve

Anterior inferior cerebellar artery

Anteromedian pontine vein

Right vertebral artery

Hypoglossal nerve

Posterior inferior cerebellar artery

Anterior spinal arteries

Anterior median fissure

Hypoglossal nerve

Medullary arteries

Anteromedian medullary vein

Anterior spinal artery

CN VI

Pyramid

(A)

(B)

(C)

Pial arterial plexus

Central (sulcal) arteries

Posterior segmental medullary arteries

Anterior segmental medullary artery

Anterior spinal artery

Posterior spinal arteries

FIGURE 8.7 (A) Arteries, prepared with India ink–gelatin, on the anterior surface of the medulla. (B–C) Cleared specimens of arteries, injected with India ink–gelatin, in relation to the (B) thoracic segment of the spinal cord, and (C) midsagittal section of the spinal cord.

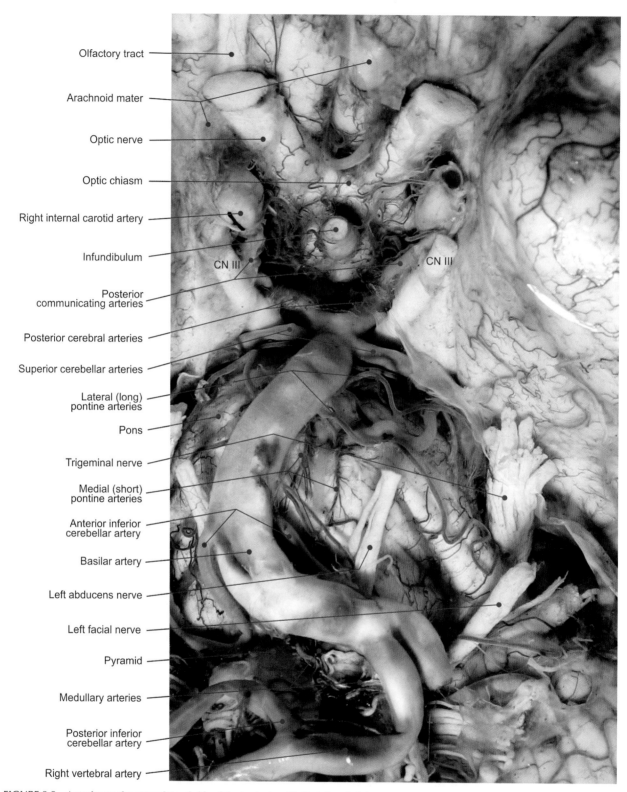

Olfactory tract

Arachnoid mater

Optic nerve

Optic chiasm

Right internal carotid artery

Infundibulum

CN III

CN III

Posterior communicating arteries

Posterior cerebral arteries

Superior cerebellar arteries

Lateral (long) pontine arteries

Pons

Trigeminal nerve

Medial (short) pontine arteries

Anterior inferior cerebellar artery

Basilar artery

Left abducens nerve

Left facial nerve

Pyramid

Medullary arteries

Posterior inferior cerebellar artery

Right vertebral artery

FIGURE 8.8 **Arteries on the anterolateral side of the brainstem (India ink–gelatin).**

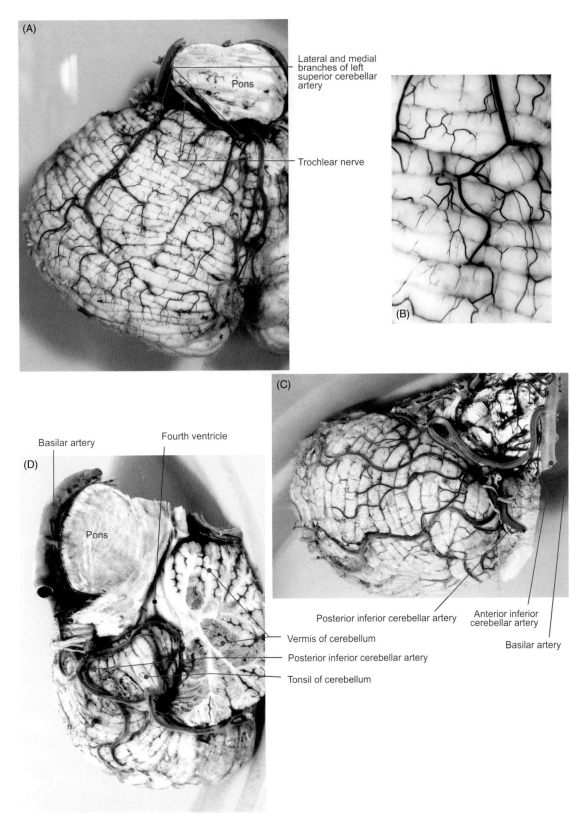

FIGURE 8.9   **The cerebellum.** (A) Vascularization of the superior aspect; (B) detail. (C) The anterior segment and (D) inferior aspect.

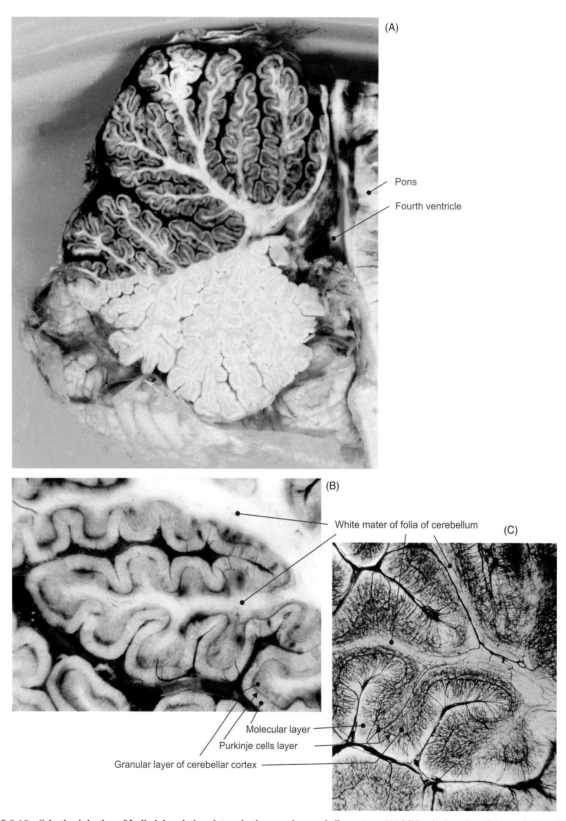

FIGURE 8.10 Selective injection of India ink–gelatin mixture in the superior cerebellar artery. (A) Midsagittal section of the cerebellum, (B) detail, and (C) cleared specimen.

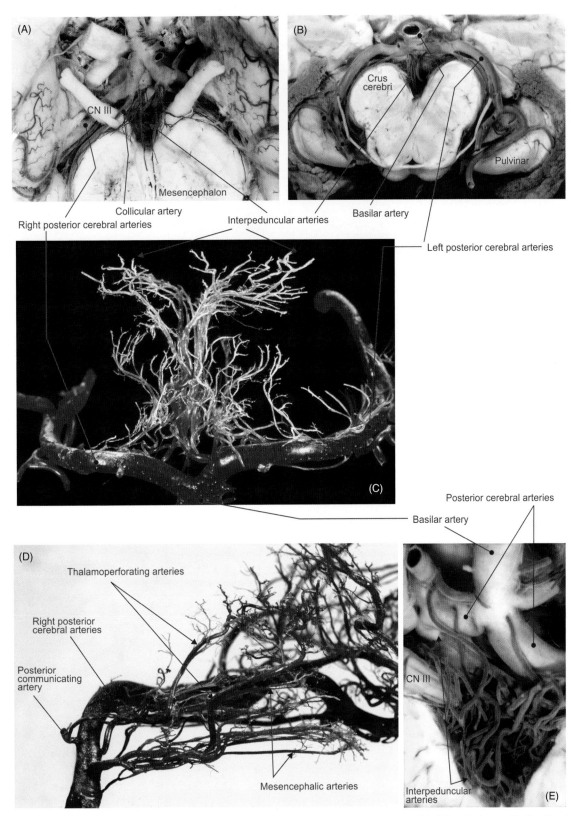

FIGURE 8.11   Interpeduncular branches of the posterior cerebral artery. (A, B, E) Basilar artery lifted from brainstem (India ink–gelatin). (C) Anterior view (corrosion cast) and (D) view from the medial side (corrosion cast).

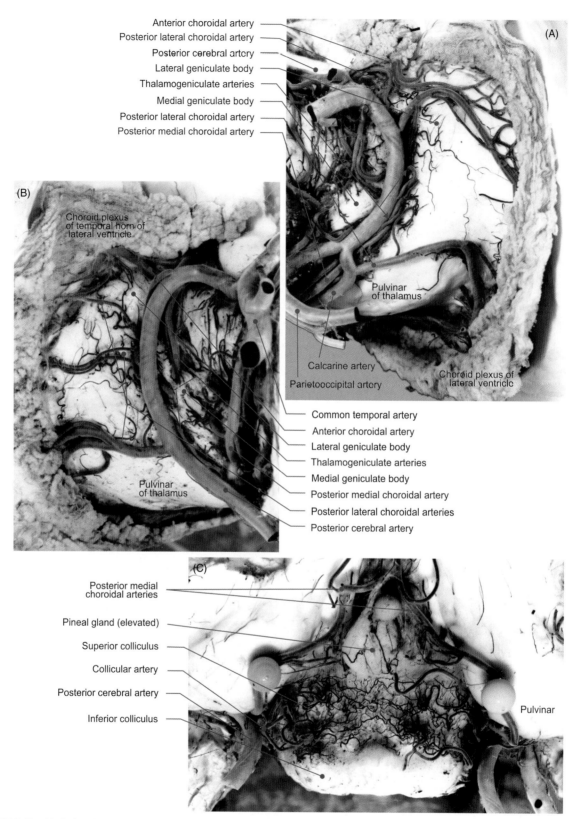

Anterior choroidal artery
Posterior lateral choroidal artery
Posterior cerebral artery
Lateral geniculate body
Thalamogeniculate arteries
Medial geniculate body
Posterior lateral choroidal artery
Posterior medial choroidal artery

(A)

Choroid plexus of temporal horn of lateral ventricle

(B)

Pulvinar of thalamus

Pulvinar of thalamus

Calcarine artery

Parietooccipital artery

Choroid plexus of lateral ventricle

Common temporal artery
Anterior choroidal artery
Lateral geniculate body
Thalamogeniculate arteries
Medial geniculate body
Posterior medial choroidal artery
Posterior lateral choroidal arteries
Posterior cerebral artery

(C)

Posterior medial choroidal arteries

Pineal gland (elevated)

Superior colliculus

Collicular artery

Posterior cerebral artery

Inferior colliculus

Pulvinar

FIGURE 8.12 (A) Left and (B) right posterior cerebral arteries and branches (India ink–gelatin). (C) Mesencephalon: arterial network around the superior colliculi (India ink–gelatin preparation).

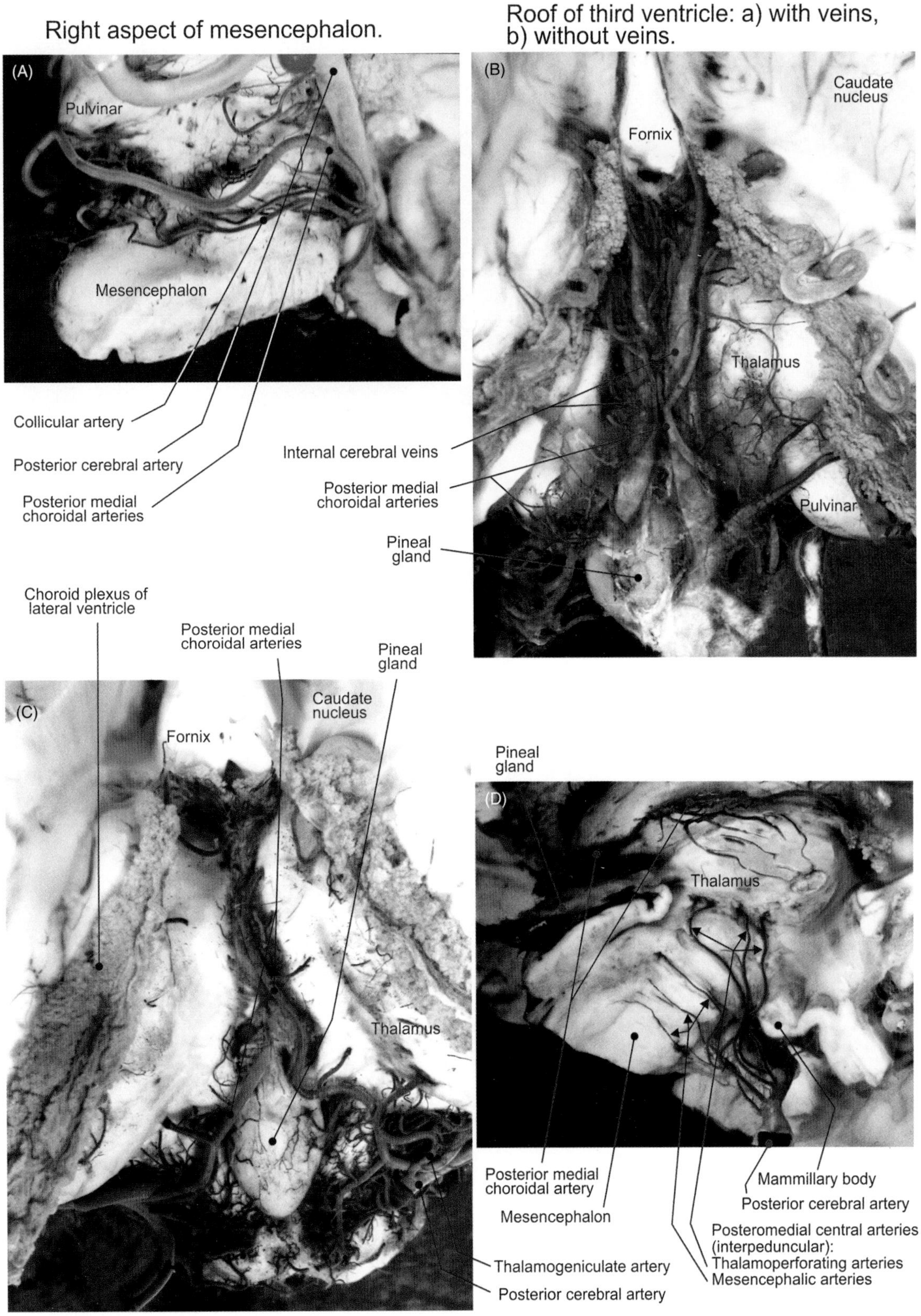

Right aspect of mesencephalon.

(A)

Pulvinar

Mesencephalon

Collicular artery

Posterior cerebral artery

Posterior medial
choroidal arteries

Roof of third ventricle: a) with veins, b) without veins.

(B)

Caudate
nucleus

Fornix

Thalamus

Internal cerebral veins

Posterior medial
choroidal arteries

Pineal
gland

Pulvinar

(C)

Choroid plexus of
lateral ventricle

Posterior medial
choroidal arteries

Pineal
gland

Caudate
nucleus

Fornix

Thalamus

(D)

Pineal
gland

Thalamus

Posterior medial
choroidal artery

Mesencephalon

Thalamogeniculate artery

Posterior cerebral artery

Mammillary body

Posterior cerebral artery

Posteromedial central arteries
(interpeduncular):
Thalamoperforating arteries
Mesencephalic arteries

FIGURE 8.13   (A) Right aspect of the mesencephalon. (B–C) Roof of the third ventricle: (B) with veins and (C) without veins. (D) Midsagittal section of the brainstem.

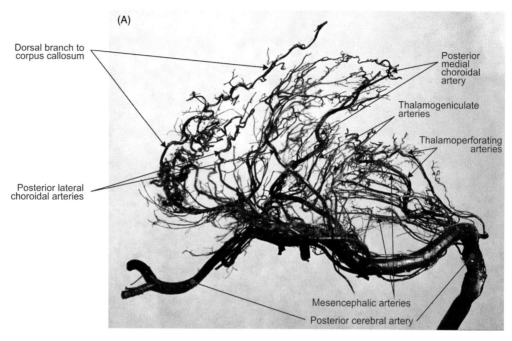

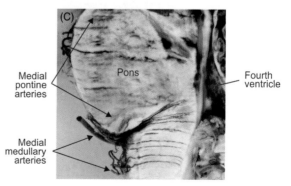

FIGURE 8.14 Deep branches of the posterior cerebral artery: view of the right artery from (A) outside and (B) inside (corrosion casts). (C) Mediosagittal section of brainstem (India ink–gelatin preparation).

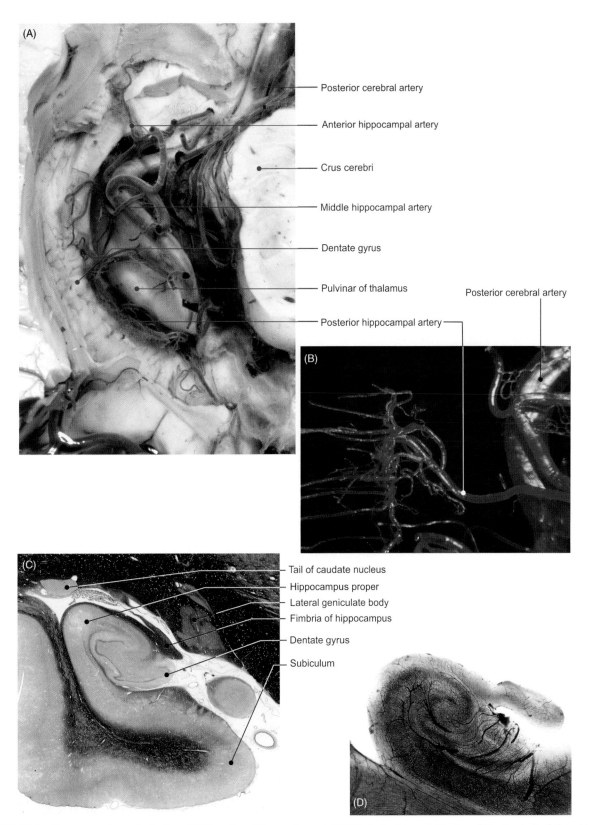

— Posterior cerebral artery

— Anterior hippocampal artery

— Crus cerebri

— Middle hippocampal artery

— Dentate gyrus

— Pulvinar of thalamus

— Posterior hippocampal artery

Posterior cerebral artery

— Tail of caudate nucleus
— Hippocampus proper
— Lateral geniculate body
— Fimbria of hippocampus
— Dentate gyrus
— Subiculum

FIGURE 8.15 **Hippocampal vascularization.** (A) Dissection, (B) corrosion cast, (C) histologic slide, and (D) view of blood vessels of cleared transverse section.

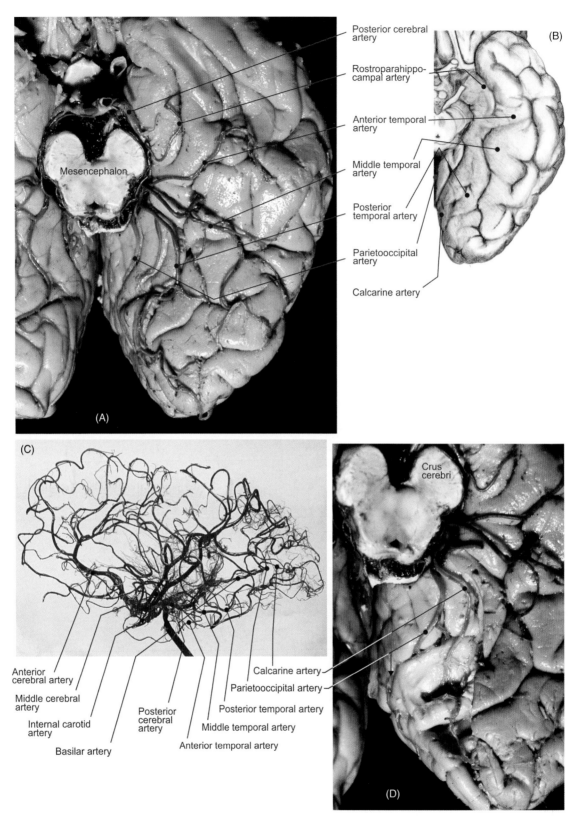

FIGURE 8.16 Cortical branches of the posterior cerebral artery: (A) India ink–gelatin preparation and (B) illustration. (C) Cortical branches of the cerebral arteries, view from outside (corrosion cast). (D) Calcarine sulcus open.

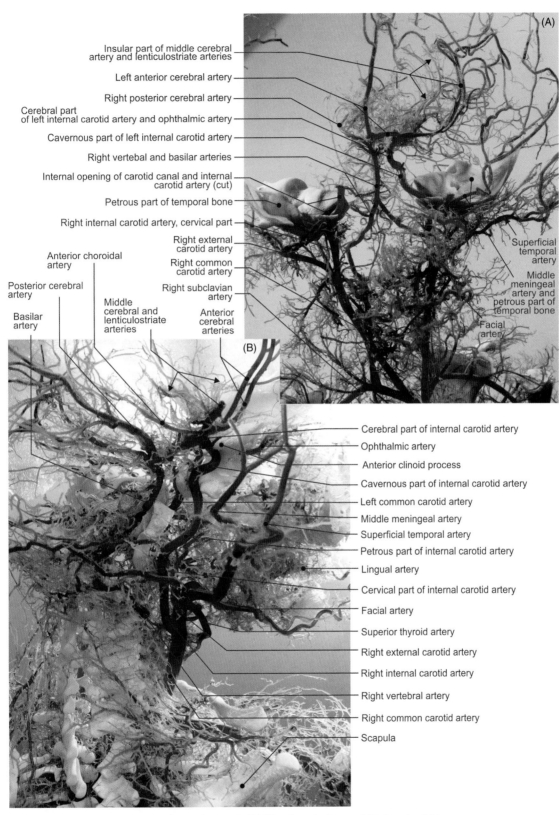

Insular part of middle cerebral artery and lenticulostriate arteries

Left anterior cerebral artery

Right posterior cerebral artery

Cerebral part of left internal carotid artery and ophthalmic artery

Cavernous part of left internal carotid artery

Right vertebal and basilar arteries

Internal opening of carotid canal and internal carotid artery (cut)

Petrous part of temporal bone

Right internal carotid artery, cervical part

Right external carotid artery

Right common carotid artery

Right subclavian artery

Anterior choroidal artery

Posterior cerebral artery

Basilar artery

Middle cerebral and lenticulostriate arteries

Anterior cerebral arteries

(A)

(B)

Superficial temporal artery

Middle meningeal artery and petrous part of temporal bone

Facial artery

Cerebral part of internal carotid artery

Ophthalmic artery

Anterior clinoid process

Cavernous part of internal carotid artery

Left common carotid artery

Middle meningeal artery

Superficial temporal artery

Petrous part of internal carotid artery

Lingual artery

Cervical part of internal carotid artery

Facial artery

Superior thyroid artery

Right external carotid artery

Right internal carotid artery

Right vertebral artery

Right common carotid artery

Scapula

FIGURE 8.17   Fetal internal carotid artery (corrosion casts). (A) View from the front and (b) from the right.

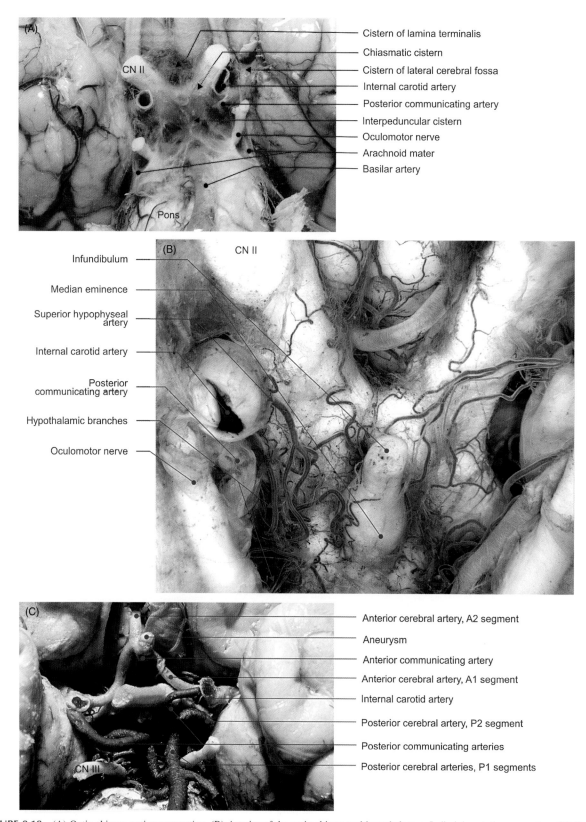

FIGURE 8.18 (A) Optic chiasm, native preparation. (B) Arteries of the optic chiasm and hypothalamus, India ink–gelatin preparation. (C) Cerebral arterial circle of Willis, silicone.

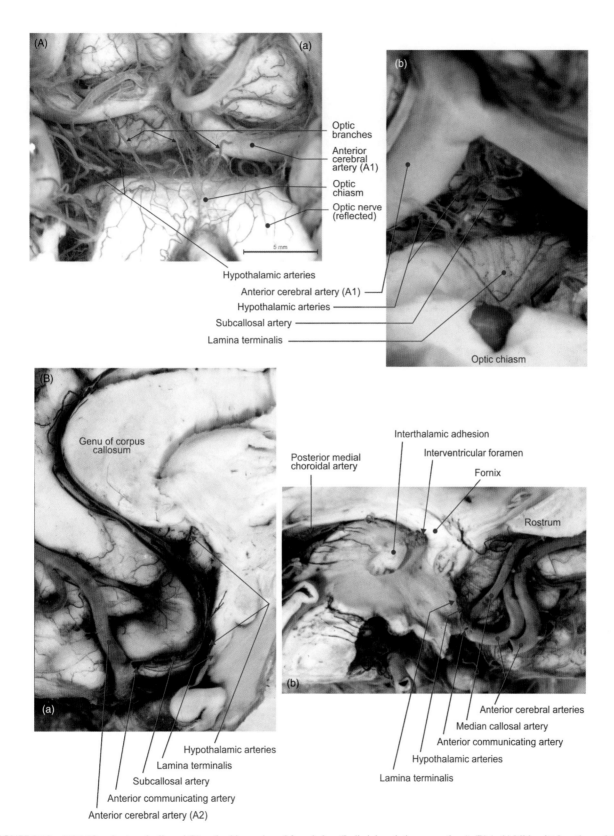

FIGURE 8.19   (A) (a) Lamina terminalis and (b) optic chiasm viewed from below (India ink–gelatin preparations). (B) (a–b) Midsagittal sections of the lamina terminalis (India ink–gelatin injection).

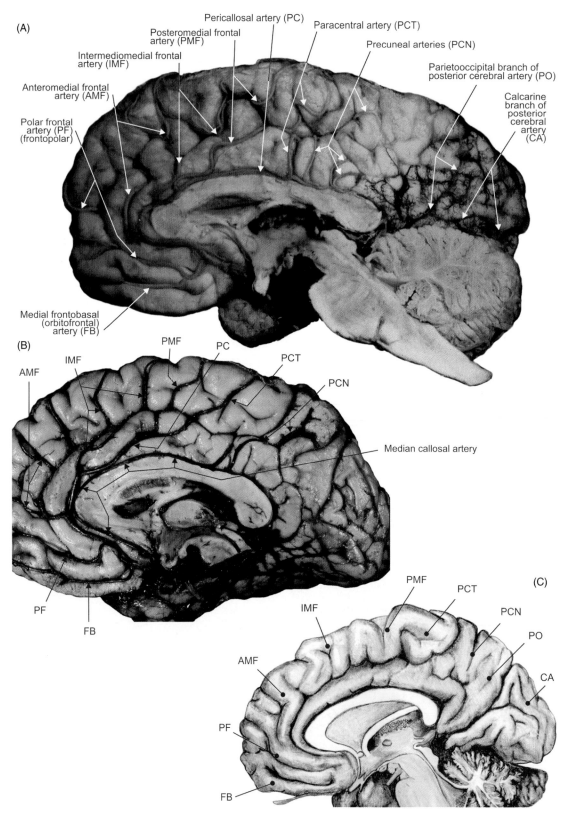

FIGURE 8.20 Cortical branches of the anterior cerebral artery on the medial aspect of the right cerebral hemispheres (India ink–gelatin preparations). (A–B) Dissections and (C) illustration.

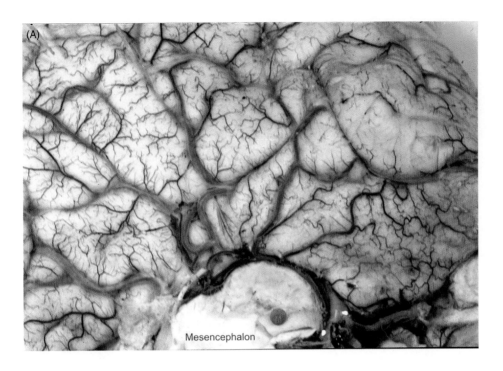

Mesencephalon

Subarachnoid space

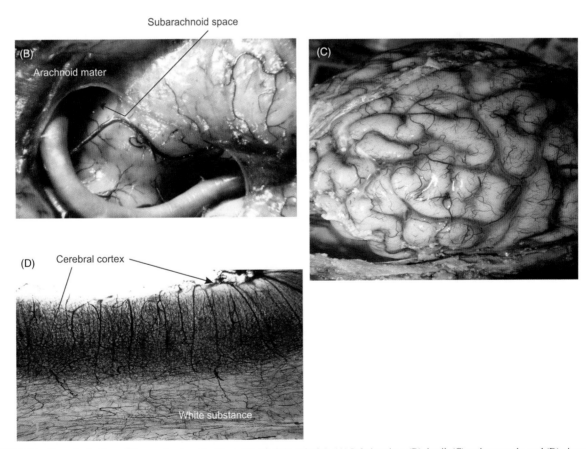

Arachnoid mater

Cerebral cortex

White substance

FIGURE 8.21   Vascularization of the cerebral cortical layer (India ink–gelatin). (A) Inferior view, (B) detail, (C) native sample, and (D) view of the cleared cortical arteries.

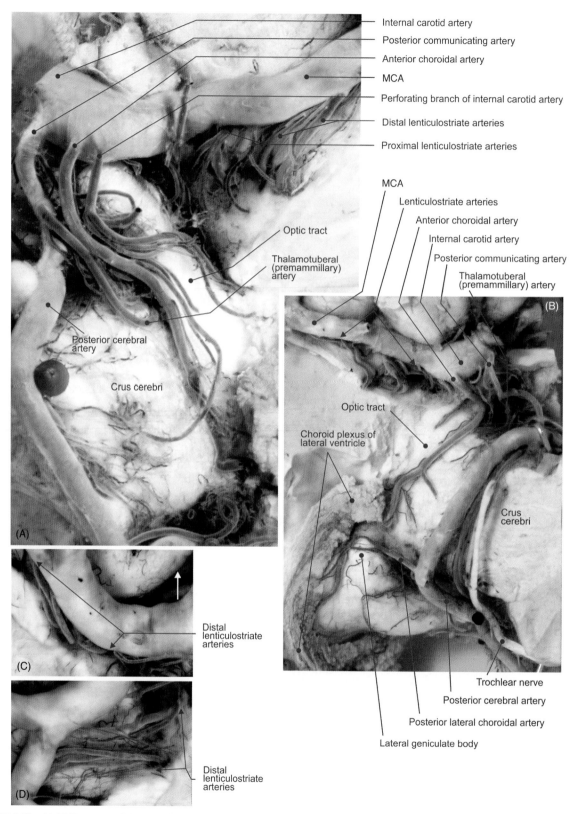

Internal carotid artery
Posterior communicating artery
Anterior choroidal artery
MCA
Perforating branch of internal carotid artery
Distal lenticulostriate arteries
Proximal lenticulostriate arteries

MCA
Lenticulostriate arteries
Anterior choroidal artery
Internal carotid artery
Posterior communicating artery
Thalamotuberal (premammillary) artery

Optic tract

Thalamotuberal (premammillary) artery

Posterior cerebral artery

Crus cerebri

(A)

(B)

Optic tract

Choroid plexus of lateral ventricle

Crus cerebri

Distal lenticulostriate arteries

(C)

Trochlear nerve
Posterior cerebral artery
Posterior lateral choroidal artery
Lateral geniculate body

Distal lenticulostriate arteries

(D)

FIGURE 8.22 (A–B) Deep part of the (A) left and (B) right hemispheres after the removal of the temporal lobe. (C–D) Trunk of the middle cerebral artery (MCA) (M1 segment), view from below: (C) before and (D) after reflection of the artery.

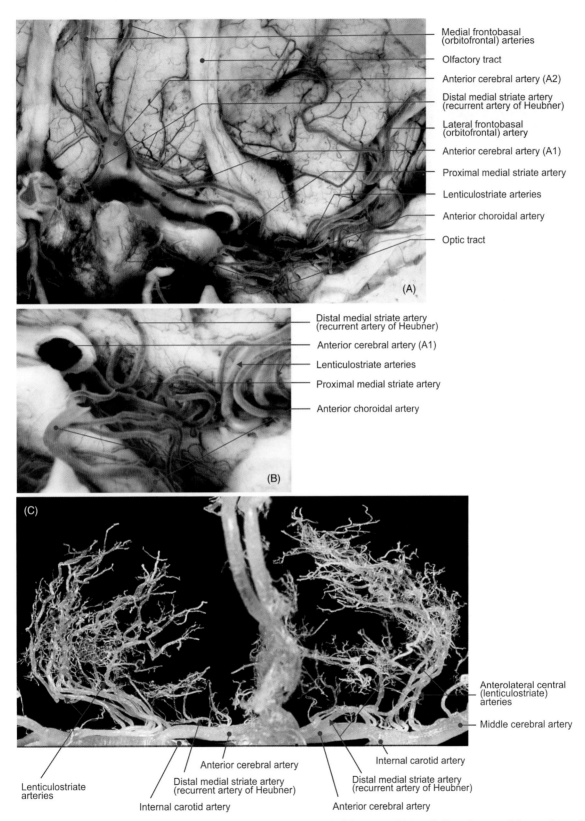

Medial frontobasal
(orbitofrontal) arteries

Olfactory tract

Anterior cerebral artery (A2)

Distal medial striate artery
(recurrent artery of Heubner)

Lateral frontobasal
(orbitofrontal) artery

Anterior cerebral artery (A1)

Proximal medial striate artery

Lenticulostriate arteries

Anterior choroidal artery

Optic tract

(A)

Distal medial striate artery
(recurrent artery of Heubner)

Anterior cerebral artery (A1)

Lenticulostriate arteries

Proximal medial striate artery

Anterior choroidal artery

(B)

(C)

Anterolateral central
(lenticulostriate)
arteries

Middle cerebral artery

Lenticulostriate
arteries

Anterior cerebral artery

Distal medial striate artery
(recurrent artery of Heubner)

Internal carotid artery

Internal carotid artery

Distal medial striate artery
(recurrent artery of Heubner)

Anterior cerebral artery

FIGURE 8.23   (A) Anterior perforated substance and (B) details, after the removal of the temporal lobe. (C) Corrosion cast of the anterior perforating arteries, view from the front.

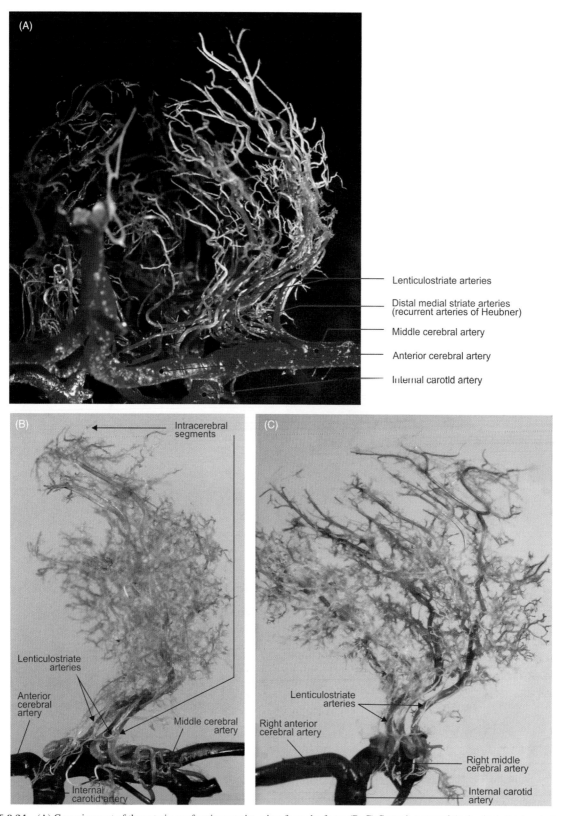

Lenticulostriate arteries

Distal medial striate arteries
(recurrent arteries of Heubner)

Middle cerebral artery

Anterior cerebral artery

Internal carotid artery

Intracerebral
segments

Lenticulostriate
arteries

Anterior
cerebral
artery

Middle cerebral
artery

Internal
carotid artery

Lenticulostriate
arteries

Right anterior
cerebral artery

Right middle
cerebral artery

Internal carotid
artery

FIGURE 8.24 (A) Corrosion cast of the anterior perforating arteries, view from the front. (B–C) Corrosion cast of the lenticulostriate arteries viewed from (B) behind and (C) medial side.

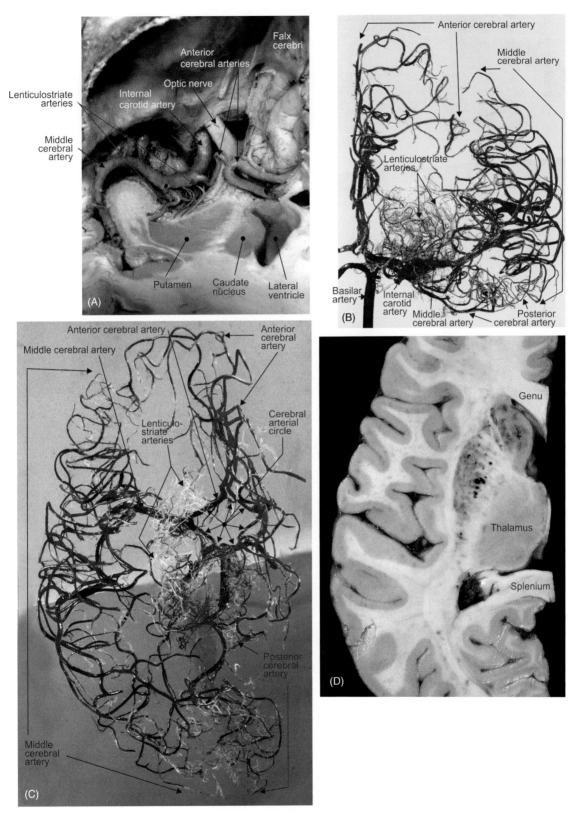

FIGURE 8.25   Left lenticulostriate arteries. (A) View from above after the removal of the frontal lobe (corrosion cast), (B) view from the front, (C) from above (corrosion cast), and (D) the transverse section of hemisphere after selective injection of arteries with India ink–gelatin mixture.

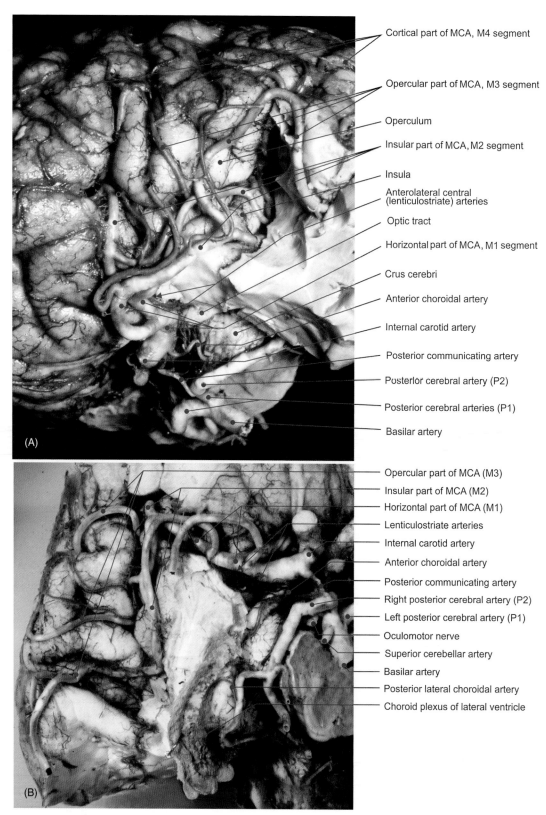

Cortical part of MCA, M4 segment

Opercular part of MCA, M3 segment

Operculum

Insular part of MCA, M2 segment

Insula

Anterolateral central (lenticulostriate) arteries

Optic tract

Horizontal part of MCA, M1 segment

Crus cerebri

Anterior choroidal artery

Internal carotid artery

Posterior communicating artery

Posterior cerebral artery (P2)

Posterior cerebral arteries (P1)

Basilar artery

(A)

Opercular part of MCA (M3)

Insular part of MCA (M2)

Horizontal part of MCA (M1)

Lenticulostriate arteries

Internal carotid artery

Anterior choroidal artery

Posterior communicating artery

Right posterior cerebral artery (P2)

Left posterior cerebral artery (P1)

Oculomotor nerve

Superior cerebellar artery

Basilar artery

Posterior lateral choroidal artery

Choroid plexus of lateral ventricle

(B)

FIGURE 8.26 Pathway and branching of the (A) left and (B) right middle cerebral arteries (MCA) after the removal of the temporal lobes (India ink–gelatin).

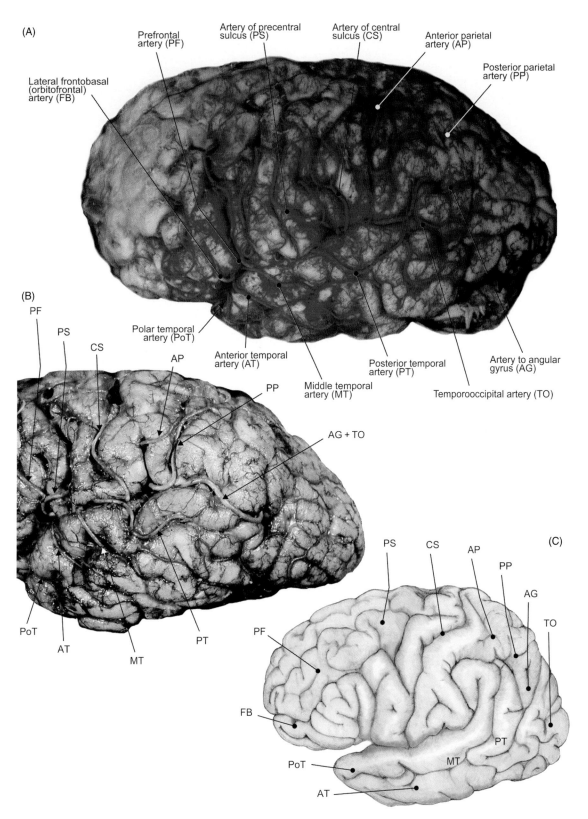

FIGURE 8.27   Cortical branches of the middle cerebral artery (India ink–gelatin): view of the superolateral aspect of the left hemisphere. (A–B) Dissections and (C) illustration.

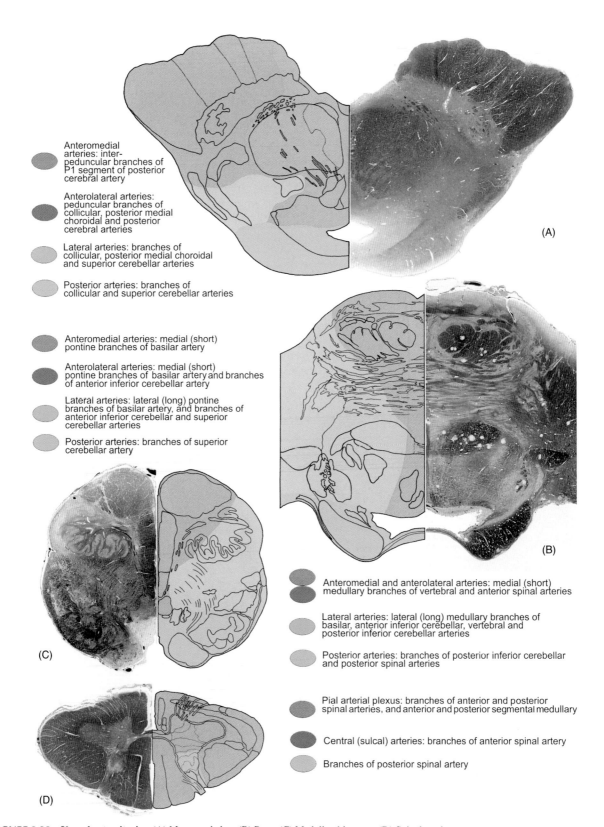

Anteromedial arteries: inter-peduncular branches of P1 segment of posterior cerebral artery

Anterolateral arteries: peduncular branches of collicular, posterior medial choroidal and posterior cerebral arteries

Lateral arteries: branches of collicular, posterior medial choroidal and superior cerebellar arteries

Posterior arteries: branches of collicular and superior cerebellar arteries

Anteromedial arteries: medial (short) pontine branches of basilar artery

Anterolateral arteries: medial (short) pontine branches of basilar artery and branches of anterior inferior cerebellar artery

Lateral arteries: lateral (long) pontine branches of basilar artery, and branches of anterior inferior cerebellar and superior cerebellar arteries

Posterior arteries: branches of superior cerebellar artery

(A)

(B)

Anteromedial and anterolateral arteries: medial (short) medullary branches of vertebral and anterior spinal arteries

Lateral arteries: lateral (long) medullary branches of basilar, anterior inferior cerebellar, vertebral and posterior inferior cerebellar arteries

Posterior arteries: branches of posterior inferior cerebellar and posterior spinal arteries

Pial arterial plexus: branches of anterior and posterior spinal arteries, and anterior and posterior segmental medullary

Central (sulcal) arteries: branches of anterior spinal artery

Branches of posterior spinal artery

(C)

(D)

FIGURE 8.28 Vascular territories. (A) Mesencephalon. (B) Pons. (C) Medulla oblongata. (D) Spinal cord.

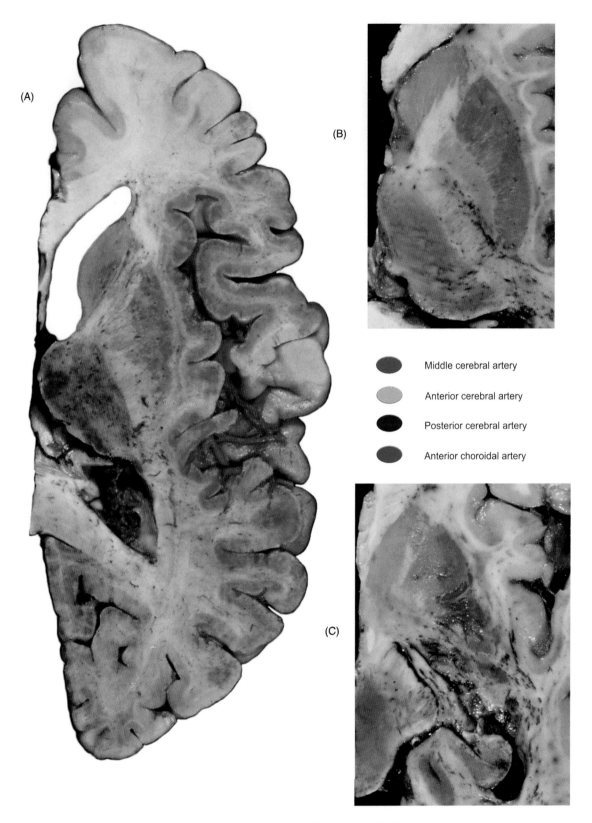

(A)

(B)

Middle cerebral artery

Anterior cerebral artery

Posterior cerebral artery

Anterior choroidal artery

(C)

FIGURE 8.29   Vascular territories on the transverse sections of the cerebral hemisphere. (A) The level of dorsal thalamus, (B) midlevel of thalamus, and (C) level ventral of thalamus.

References

Abrahams, P., Spratt, J., Loukas, M., van Schoor, A.-N., 2013. McMinn & Abrahams' Clinical Atlas of Human Anatomy. Elsevier Mosby; London.

Carpenter, M., 1991. Core Text of Neuroanatomy. Williams & Wilkins, Baltimore; Tokyo.

Ćetković, M., Antunović, V., et al., 2011. Vasculature and neurovascular relationships of the trigeminal nerve root. Acta Neurochir. 153 (5), 1051–1057.

DeArmond, S., Fusco, M., Dewey, M., 1976. Structure of the Human Brain. University Press, New York; Oxford.

Drake, R., Wayne Vogl, A., Mitchell, A., 2015. Gray's Anatomy. Churchill Livingstone Elsevier, Philadelphia.

Duvernoy, H., 1978. Human Brainstem Vessels. Springer Verlag, Berlin; Heidelberg; New York.

Duvernoy, H., 1985. The Human Hippocampus. JF Bergmann Verlag, Munchen.

Ellis, H., Logan, B., Dixon, A., 1999. Human Sectional Anatomy. Atlas of Body Sections, CT and MRI Images. Butterworth–Heinemann, Oxford; New Delhi.

Gambarelli, J., Guerinel, G., Chevrot, L., Mattei, M., 1977. Computerized Axial Tomography. An Anatomic Atlas of Sections of the Human Body. Springer Verlag, Berlin; New York.

Gerhardt, P., 1988. Atlas of Anatomic Correlations in CT and MRI. Thieme Verlag, Stuttgart; New York.

Gilroy, A., MacPherson, B., Schuenke, M., Schulte, E., Schumacher, U., 2016. Atlas of Anatomy. Thieme, New York; Stuttgart; Delhi; Rio de Janeiro.

Hagens, G., Romrell, L., Ross, M., Tiedemann, K., 1990. Visible Human Body. Lea & Febiger, Philadelphia; London.

Haines, D., 2011. Neuroanatomy: An Atlas of Structures, Sections, and Systems. Lippincott Williams & Wilkins, Baltimore.

Hanaway, J., Woolsey, T., Gado, M., Roberts, M., 1998. The Brain Atlas. Fitzgerald Science Press, Inc., Bethesda.

Koritke, J., Sick, H., 1983. Atlas of Sectional Human Anatomy. Urban & Schwarzenberg, Baltimore; Munich.

Mai, J., Assheuer, J., Paxinos, G., 1997. Atlas of the Human Brain. Academic Press, San Diego; Toronto.

Marinković, S., Gibo, H., Milisavljević, M., et al., 2001. Anatomic and clinical correlations of the lenticulostriate arteries. Clin. Anat. 14 (3), 190–195.

Marinković, S., Gibo, H., Milisavljević, M., et al., 2005. Microanatomy of the intrachoroidal vasculature of the lateral ventricle. Neurosurgery 57 (Suppl. 1), 22–36.

Marinković, S., Milisavljević, M., et al., 1985. Perforating branches of the middle cerebral artery. Microanatomy and clinical significance of their intracerebral segments. Stroke 16, 1022–1029.

Marinković, S., Milisavljević, M., et al., 1986a. Anastomoses among the thalamoperforating branches of the posterior cerebral artery. Arch. Neurol. 43, 811–814.

Marinković, S., Milisavljević, M., et al., 1986b. Anatomical bases for surgical approach to the initial segment of the anterior cerebral artery. Microanatomy of Heubner's artery and perforating branches of the anterior cerebral artery. Surg. Radiol. Anat. 8, 7–18.

Marinković, S., Milisavljević, M., et al., 1986c. Interpeduncular perforating branches of the posterior cerebral artery. Microsurgical anatomy of their extracerebral and intracerebral segments. Surg. Neurol. 26, 349–359.

Marinković, S., Milisavljević, M., et al., 1987. Distribution of the occipital branches of the posterior cerebral artery. Correlation with occipital lobe infarcts. Stroke 18, 728–732.

Marinković, S., Milisavljević, M., et al., 1990a. Branches of the anterior communicating artery. Microsurgical anatomy. Acta. Neurochir. 106 (1–2), 78–85.

Marinković, S., Milisavljević, M., et al., 1990b. The perforating branches of the internal carotid artery: the microsurgical anatomy of their extracerebral segments. Neurosurgery 26 (3), 472–478.

Marinković, S., Milisavljević, M., et al., 1992. Microvascular anatomy of the hippocampal formation. Surg. Neurol. 37, 339–349.

Marinković, S., Milisavljević, M., et al., 2004. Microsurgical anatomy of the perforating branches of the vertebral artery. Surg. Neurol. 61 (2), 190–197.

Marinković, S., Schellinger, D., Milisavljević, M., Antunović, V., Petrović, P., Vidić, B., Gibo, H., Štimec, B., Maliković, A., 2000. Sectional and MRI Anatomy of the Human Body. CIC Edizioni Internazionali; Roma.

Maliković, A., Vučetić, B., Milisavljević, M., et al., 2012. Occipital sulci of the human brain: variability and morphometry. Anat. Sci. Int. 87 (2), 61–70.

Martin, J., 1989. Neuroanatomy, Text and Atlas. Elsevier, New York; London.

Milisavljević, M., Marinković, S., et al., 1986. Anastomoses in territory of the posterior cerebral artery. Acta Anat. 127, 221–225.

Milisavljević, M., Marinković, S., et al., 1988. Anatomic basis for surgical approach to the distal segment of the posterior cerebral artery. Surg. Radiol. Anat. 10, 259–266.

Milisavljević, M., Marinković, S., et al., 2013. Duplication of the superior vena cava associated with atrial termination of the left hepatic vein. Phlebology 28 (7), 369–374.

Milisavljević, M., Marinković, S., et al., 1986b. Oculomotor, trochlear and abducens nerves penetrated by cerebral vessels. Arch. Neurol. 43, 58–61.

Milisavljević, M., Marinković, S., et al., 1991. The thalamogeniculate perforators of the posterior cerebral artery—the microsurgical anatomy. Neurosurgery 28 (4), 523–530.

Milisavljević, M., Vitošević, Z., Vidić, B., Maliković, A., 2011. Atlas Disekcije Čoveka. Data Status, Beograd.

Mtui, E., Gruener, G., Dockery, P., 2016. Fitzgerald's Clinical Neuroanatomy and Neuroscience. Elsevier, Philadelphia.

Netter, F., 2014. Atlas of Human Anatomy. Saunders & Elsevier, Philadelphia.

Nolte, J., Angevine, J., 1995. The Human Brain. In Photographs and Diagrams. Mosby, St. Louis; Wiesbaden.

Paulsen, F., Waschke, J., 2013. Sobotta Atlas of Human Anatomy. Elsevier, Munich.

Roberts, M., Hanaway, J., Morest, K., 1987. Atlas of the Human Brain in Sections. Lea & Febiger, Philadelphia.

Rohen, J., Yokochi, C., Lutjen-Drecoll, 1998. Color Atlas of Anatomy. Williams & Wilkins, Baltimore.

Whitmore, I., Federative Committee on Anatomical Terminology, 1998. Terminologia Anatomica. Thieme, Stuttgart; New York.

Toldt, C., Dalla Rosa, A., 1948. An Atlas of Human Anatomy. Macmillan Company, New York.

Vidić, B., Suarez, F., 1984. Photographic Atlas of the Human Body. Mosby, St Louis; Toronto.

Vitošević, Z., Ćetković, M., et al., 2005. Vaskularizacija kapsule interne i bazalnih jedara. Srp. Arh. Celok. Lek. 133 (1–2), 41–45.

Vitošević, Z., Marinković, S., Ćetković, M., Štimec, B., et al., 2013. Intramesencephalic course of the oculomotor nerve fibers: microanatomy and possible clinical significance. Anat. Sci. Int. 88 (2), 70–82.

Watson, C., 1985. Basic Human Neuroanatomy: An Introductory Atlas. Little, Brown and Company, Boston; Toronto.

Yasargil, M., 1984. Microneurosurgery. Georg Thieme Verlag, Stuttgart; New York.

Zuleger, S., Staubesand, J., 1977. Atlas of the Central Nervous System in Sectional Planes. Urban & Schwarzenberg, Baltimore; Munich.

Index